AGRICULTURE ISSUES AND POLICIES

NITROGEN FERTILIZER

AGRICULTURAL USES, MANAGEMENT PRACTICES AND ENVIRONMENTAL EFFECTS

AGRICULTURE ISSUES AND POLICIES

Additional books in this series can be found on Nova's website under the Series tab.

Additional e-books in this series can be found on Nova's website under the e-book tab.

AGRICULTURE ISSUES AND POLICIES

NITROGEN FERTILIZER

AGRICULTURAL USES, MANAGEMENT PRACTICES AND ENVIRONMENTAL EFFECTS

KARLENE WINSLOW
EDITOR

nova publishers
New York

For permission to use material from this book please contact us:
Telephone 631-231-7269; Fax 631-231-8175
Web Site: http://www.novapublishers.com

Additional color graphics may be available in the e-book version of this book.

Library of Congress Cataloging-in-Publication Data

Nitrogen fertilizer : agricultural uses, management practices and environmental effects / editor: Karlene Winslow.
pages cm. -- (Agriculture issues and policies)
Includes index.
ISBN 978-1-63321-585-6 (hardcover)
1. Nitrogen fertilizers. 2. Nitrogen fertilizers--Environmental aspects. I. Winslow, Karlene. II. Series: Agriculture issues and policies series.
S651.N56 2014
631.8'4--dc23
2014027189

Published by Nova Science Publishers, Inc. † New York

CONTENTS

PREFACE

Nitrate and nitrite are two ions largely diffused in the environment because they take part in the nitrogen cycle. Moreover, a great part of atmospheric nitrogen may be oxidized to nitrite and nitrate by microorganisms in plants, soil or water. The more stable form of oxidized nitrogen is nitrate ion, but, through microbial activity, it can be reduced to nitrite ion which is more chemically reactive. Nitrate and its salts are widely used, especially as inorganic fertilizers, and for many other purposes such as oxidizing agents, explosives, in the chemical industry and as food preservatives. This book discusses the agricultural uses, management practices and environmental effects of nitrogen fertilizers.

Chapter 1 – Nitrites and nitrates are well known compounds, largely used as food additives especially for meat curing. A natural occurrence of nitrates in some foodstuffs may be due to the presence of this compound in the environment as consequence of the nitrogen cycle; while nitrites may be present, but at very low concentrations.

Due to toxic effects that nitrites and nitrates may exercise on humans, the levels of these compounds in foodstuffs are continuously monitored by health authorities. In the last decade, these controls were intensified because the large use of nitrogen fertilizers may bring about an increase of the levels of these substances in several food products.

Vegetables are products characterized by extremely variable levels of nitrates, accumulated in large amounts especially in the leaves, in relationship to cultivar, light intensity, temperature, carbon dioxide level, water relation, soil type and, above all, N-fertilizer application timing.

Other foodstuffs of animal origin, such as fresh meats and dairy products, may be characterized by not negligible concentrations of nitrates; nitrites may be also present, due to endogenous reduction of nitrates. These amounts are due to nitrates levels in animal feeds that may reach concentrations up to 1760 mg kg^{-1} in fresh alfalfa.

Seafood, in particular shellfish, may also accumulate nitrates as consequence of nitrogen pollution both in groundwater and surface water.

In this chapter the results of several surveys, focused on tracing quantifiable amounts of nitrites and nitrates in the most important foodstuffs, such as fresh meats, dairy products, seafood, leafy vegetables and in animal feeds, are reported.

The results showed that nitrate levels may be very high in leafy vegetables, where concentrations higher than legal limits were registered. Concentrations higher than limit of quantification (9.6 mg kg^{-1}) where detected in some types of fresh meats and of dairy products; while in mussels these levels may reach 400 mg kg^{-1}. Nitrites are present at low concentrations; however, high concentrations were detected in some leafy vegetables, up to 197.5 mg kg^{-1}.

In conclusion, the results suggest the introduction of specific limits for some contaminant/matrix combinations and demonstrate the importance of official controls.

Chapter 2 – The demand for urea is continuously growing and entwined with the need for fertilizers and animal feed additives.

Industrially, urea is initially produced in liquid form as a concentrated solution. Then, it can be converted into particulate material either through granulation or prilling processes. Since granules have better attributes than prills, nowadays granulation is the preferred production route. Urea granulation is a multifaceted process that requires several operation units, which constitute the granulation circuit, to produce the solid form (granules) with the desired attributes. The main unit of the circuit is the granulator, where small urea particles known as seeds are continuously fed and sprayed with a urea concentrated solution. The seeds grow through deposition of the solution droplets onto the solids surface, followed by water evaporation and urea solidification. The granules that leave the size enlargement unit are size classified into product, oversize and undersize streams. The product is transported to storage facilities, while the oversize fraction is fed to crushers for size reduction. The crushed oversize particles are then combined with the undersize granules and recycled back to the granulator as seeds.

Focusing on urea, the advantages of granules over prills are discussed by exploring the physical properties of both solid forms. Then, the current available technologies for urea granulation are presented in a comparative manner. From this analysis, the fluidized-bed granulator appears as the most widely used equipment for granular urea production. Due to this preference, different approaches to model fluidized-bed granulators are presented aiming to give a comprehensive picture of the fundamental phenomena that occurs within these granulation units. Special attention is placed on the granules growth mechanism, and its proper representation. Although coating is the preferred urea growth mechanism, unexpected operating situations may favor size enlargement by agglomeration that is an undesired phenomenon. Therefore, based on experimental data obtained in a pilot-scale fluidized-bed batch granulator for urea production, the influence of the operating variables on both granules quality and growth mechanisms is discussed. Finally, mathematical models for peripheral circuit units (crusher, cooler and screens) are presented. By coupling all the involved units, a complete granulation circuit simulator is reported. Steady-state and dynamics results obtained by means of the urea granulation simulator are provided to show the influence of different circuit operating variables on the marketable product size distribution and the plant throughput. Summarizing, this chapter gives an introduction to the main features of the urea granulation process and remarks operation problems that face the granular urea production together with possible strategies to overcome them.

Chapter 3 – The excessive use of mineral fertilizers increases the production costs of agricultural crops, while also causing the acidification of soil and deterioration of chemical and microbiological properties. In the case of mineral nitrogenous fertilizers the soil is very unstable and is easily washed off so that eutrophication of groundwater occurs. The aim of this study was to investigate the possibility of reduction, as well as the complete exclusion, of mineral nitrogen fertilizer in soybean production through the application of: symbiotic bacteria *Bradyrhizobium japonicum*, associative bacteria *Azospirillum brasilense* and non symbiotic bacteria *Azotobacter chroococcum*. Reduction of nitrogen fertilizer through nitrogen-fixing bacteria substitution doesn't only have an econo-mic importance in agricultural production, but is also of great importance to the ecology of the environment.

During two-year investigations on two different soil types (Humogley, Eutric Cambisols) in 4 repetitions and 9 different variants, as follows: 1. control (untreated seed + mineral nitrogen fertilization based on soil analysis); 2. seed treated with symbiotic bacteria *B. japonicum* + mineral nitrogen fertilization based on soil analysis; 3. seed treated with symbiotic bacteria *B. japonicum* + reduced mineral nitrogen fertilization by 30%; 4. seed treated with symbiotic bacteria *B. japonicum* + reduced mineral nitrogen fertilization by 50%; 5. seed treated with symbiotic bacteria *B. japonicum* + reduced mineral nitrogen fertilization by 70%; 6. seed treated with symbiotic bacteria *B. japonicum,* associative bacteria *A. brasilense* and non symbiotic bacteria *A. chroococcum* + mineral nitrogen fertilization based on soil analysis; 7. seed treated with symbiotic bacteria *B. japonicum,* associative bacteria *A. brasilense* and non symbiotic bacteria *A. chroococcum* + reduced mineral nitrogen fertilization by 30%; 8. seed treated with symbiotic bacteria *B. japonicum,* associative bacteria *A. brasilense* and non symbiotic bacteria *A. chroococcum* + reduced mineral nitrogen fertilization by 50%; 9. seed treated with symbiotic bacteria *B. japonicum,* associative bacteria *A. brasilense* and non symbiotic bacteria *A. chroococcum* + reduced mineral nitrogen fertilization by 70%.

All seed inoculated variants accomplished significantly higher results in comparison with control variants, on both soil types and in both years of investigation.

In both years of investigation on Humogley, a soil type with greater pedological, physical, chemical and microbiological characteristics, the best results of investigated parameters were obtained in the seed variant inoculated with all three different species of benefit bacteria (*B. japonicum, A. brasilense, A. chroococcum*) and with reduced mineral nitrogen fertilization by 70%.

On Eutric Cambisols, in the first year of investigation, best results of investigated parameters were obtained in the seed variant inoculated with all three different species of benefit bacteria and with reduced mineral nitrogen fertilization by 50%. In the second year of investigation, which was marked by a lack of rainfall during July and September, at individual investigated parameters there were no statistically significant differences between listed fertilization and fertilization reduced by 30%.

Chapter 4 – In developing countries the supply of cash crops of nitrogen is usually based on unfounded rates and did not successfully change–with fertilization, based on N analyses and calculated nitrogen target values. In Slovenia, the first analyses of the sap test in winter wheat were done 30 years ago, but the soil Nmin analyses were not supported from university professors and even from the government until 2004. The soil Nmin sometimes reached more than 400 kg Nmin ha^{-1} up to 0.9 m soil depth, which was associated with nitrogen pollution of drinking water with over 50 mg NO_3^- l^{-1}. In the late 1990s, three researchers (Bavec, Bavec and Briški) provided analyses of soil Nmin and established nitrogen target values despite objections. Analyses are involved for practical farm use with EU support as an agri-environmental scheme after establishing integrated field crop production standards. Slovenia is the only country that provided this measure at a national level in EU 27 in the period 2004-2014. The data from research and analyses of integrated crop production praxis with a defined system of soil Nmin analyses were done in this chapter. Presentation of Nmin data and some problems of introduction of soil Nmin analyses into the support system were analyzed and represented. The analyses of Nmin, based on 5075 soil samples, also showed that the timing of sampling was sometimes incorrect, but timely analyses showed appropriate levels of Nmin in the soil profiles. However, in the new Program of rural development for 2014-2020, the nitrogen fertilization based on the soil Nmin analyses, for the Slovenian Ministry of agriculture and the environment, is a system too complicated for money transfer for economically and environmentally acceptable N supply in field crop production, which complies to EU CAP and EU Nitrate directive. However, producers, professionals, non-governmental and governmental institutions need to renew calculations of economical benefits, impacts and environmental risks for better development and reasonable N supply in production of their main cash crop. Because of discrepancies among countries, the minimal N management standards due to production costs and negative environmental impacts needs to be regulated and supported around the world by the FAO and/or OECD.

In: Nitrogen Fertilizer
Editor: Karlene Winslow

ISBN: 978-1-63321-585-6

Chapter 1

NITROGEN FERTILIZERS AND NITRITE-NITRATE ACCUMULATION IN FOODSTUFFS: A SURVEY

Marco Iammarino and Aurelia Di Taranto
Istituto Zooprofilattico Sperimentale della Puglia e della Basilicata, Foggia, Italy

ABSTRACT

Nitrites and nitrates are well known compounds, largely used as food additives especially for meat curing. A natural occurrence of nitrates in some foodstuffs may be due to the presence of this compound in the environment as consequence of the nitrogen cycle; while nitrites may be present, but at very low concentrations.

Due to toxic effects that nitrites and nitrates may exercise on humans, the levels of these compounds in foodstuffs are continuously monitored by health authorities. In the last decade, these controls were intensified because the large use of nitrogen fertilizers may bring about an increase of the levels of these substances in several food products.

Vegetables are products characterized by extremely variable levels of nitrates, accumulated in large amounts especially in the leaves, in relationship to cultivar, light intensity, temperature, carbon dioxide level, water relation, soil type and, above all, N-fertilizer application timing.

Other foodstuffs of animal origin, such as fresh meats and dairy products, may be characterized by not negligible concentrations of nitrates; nitrites may be also present, due to endogenous reduction of nitrates. These amounts are due to nitrates levels in animal feeds that may reach concentrations up to 1760 mg kg^{-1} in fresh alfalfa.

Seafood, in particular shellfish, may also accumulate nitrates as consequence of nitrogen pollution both in groundwater and surface water.

In this chapter the results of several surveys, focused on tracing quantifiable amounts of nitrites and nitrates in the most important foodstuffs, such as fresh meats, dairy products, seafood, leafy vegetables and in animal feeds, are reported.

The results showed that nitrate levels may be very high in leafy vegetables, where concentrations higher than legal limits were registered. Concentrations higher than limit of quantification (9.6 mg kg^{-1}) where detected in some types of fresh meats and of dairy products; while in mussels these levels may reach 400 mg kg^{-1}. Nitrites are present at low concentrations; however, high concentrations were detected in some leafy vegetables, up to 197.5 mg kg^{-1}.

In conclusion, the results suggest the introduction of specific limits for some contaminant/matrix combinations and demonstrate the importance of official controls.

INTRODUCTION

Sources of Nitrite and Nitrate in Foodstuffs

Nitrate and nitrite are two ions largely diffused in the environment because they take part in the nitrogen cycle. Moreover, a great part of atmospheric nitrogen may be oxidized to nitrite and nitrate by microorganisms in plants, soil or water.

The more stable form of oxidized nitrogen is nitrate ion, but, through microbial activity, it can be reduced to nitrite ion which is more chemically reactive.

Nitrate and its salts are widely used especially as inorganic fertilizers and for many other purposes such as oxidizing agents, explosives, in the chemical industry and as food preservatives.

Due to massive use of nitrogen fertilizers and to fossil fuels combustion, the accumulation of nitrate in the environment increased dramatically during the second half of the twentieth century [1-4].

An important consequence of this accumulation is an increase of the residual amounts of nitrite and nitrate in several types of foodstuff. The different aspects related to nitrate and nitrite accumulation in foodstuffs have been largely deepened since 1972.

In 1993, the Committee of experts on health control of foodstuffs of Council of Europe issued a very important document about nitrites and nitrates in foodstuffs [5]. This document represents a focus, since it examines in depth both the aspects related to nitrites and nitrates occurrence in foods and to toxicity in man and human exposure. In 1997, the Scientific Committee for Food of European Commission published a report about nitrite and nitrate. This document reported that improvements in manufacturing practice of cured products brought about a steady decline of nitrate levels over the years, with mean levels in these products in the range 10-30 mg kg^{-1}. An interesting in-depth analysis about N-nitrosocompounds, deriving from the reaction between secondary amines and nitrites in cured meat, beer and fish, was also reported. N-nitrosocompounds (volatile nitrosamines) are dangerous compounds for humans since they were associated with gastric cancer [6-12].

In June 2004, the New Zealand Food Safety Authority issued a document named "Nitrates and nitrites dietary exposure and risk assessment" in which the most important foodstuffs were monitored. This document reported that green vegetables (especially watercress, celery and lettuce) were the most contaminated by nitrates (mean concentrations range: 133-1640 mg kg^{-1}), whereas other vegetables and the most important processed foods (meats and cheeses) showed nitrates concentrations lower than 65.8 mg kg^{-1}. As it regards nitrites this monitoring reported very low concentrations (lower than 35.6 mg kg^{-1}) also for cured meats added with nitrites [13].

The United States Environmental Protection Agency published, in 2007, an exposure assessment for children's health relating to nitrite and nitrate. This text reported that the most likely exposure pathways for children are the ingestion of contaminated drinking water and of food containing preservatives, such as cured meats and hot dogs. Moreover, the U.S. EPA, in this document, states as follows: "Infants below the age of six months who drink water containing nitrate in excess of the MCL (1 mg L^{-1} for nitrites and 10 mg L^{-1} for nitrates) could become seriously ill and, if untreated, may die" [14].

In a fact sheet released in 2011 [15], the possibility to acquire methaemoglobinemia, for infants younger than 4 months of age, exposed to high levels of nitrates/nitrites as a consequence of drinking water or eating food from areas containing nitrogen-based fertilizers, was also highlighted by the Division of Toxicology and Environmental Medicine (ToxFAQs™) of U.S. Agency for toxic substances and disease registry.

Recently, the problem related to nitrate accumulation both in groundwater and surface water has gained much in importance. The use of high amounts of nitrate in organic and chemical fertilizers caused an increment of ~50% of nitrogen discharge into surface water across all Member States in Europe. Consequently, the European Community released the EU Nitrates Directive, in 1991, with the aim to promote the good farming practices for the prevention of nitrates accumulation in ground and surface waters across Europe [16]. Moreover, a nitrate level higher than the maximum admitted level of nitrate in drinking water (10 mg L^{-1} as N in USA and 50 mg L^{-1} as NO_3^- in Europe) was ascertained several times in the last years; this phenomenon was correlated with agricultural practices [17-19].

An important parameter was defined and reported in the Background document for development of World Health Organization Guidelines for Drinking-water [20]. This parameter, relating to simultaneous occurrence of nitrate and nitrite in drinking water, defines a limit for the sum of concentration (C) / guideline value (GV) ratios of nitrite and nitrate:

$$\frac{C_{nitrate}}{GV_{nitrate}} + \frac{C_{nitrite}}{GV_{nitrite}} \leq 1$$

Where the guideline values are equal to 3 and 50 mg L^{-1} for nitrite and nitrate respectively.

Other than contaminated water, animal feeds are another important source of nitrites and nitrates that may bring about an increase of the amounts of these compounds in foodstuffs.

High nitrates levels in forages are often due to drought conditions that result in poor pastures and reduced forage yields. In these cases, producers wish to use barren or low producing grain crops to replace forages. In the past, several cases of abortions caused by immature oat hay, of corn stalk poisoning and of decreased milk production were attributed to high nitrate levels in the forages.

Nitrates are a natural compound of all plants. Excessively amounts in forages may be due to stress conditions, such as: detrimental weather (low temperatures, drought, hail, frost, etc.); low light intensity; herbicide applications; diseases. Nitrates levels in plants also depend on: growth stage (high levels in young plants and decreasing with plant maturation); plant species (some plants like oat, sorghum, corn, etc. are high in nitrate, whereas other legumes and grasses may have high levels only under extreme conditions); plant part (in relationship with plant species); field location; time of light exposure a.m./p.m.; rains; nitrogen fertilization [21].

This last point, and in particular the use of high amounts of ammonium nitrate fertilizers, may cause poisoning conditions in animals, due to nitrate fraction that accumulates in the rumen and then can be converted to nitrite before crossing the rumen wall. When nitrite level in the rumen exceeds the ability of the microbes to convert it to ammonia, nitrate poisoning may occur. This is due to nitrite combination with hemoglobin, after it absorption through the rumen wall into the bloodstream, and subsequent formation of methemoglobin that is unable to carry oxygen to body tissues [22].

The Panel on Contaminants in the Food Chain of European Food Safety Authority released a Scientific Opinion titled "Nitrite as undesirable substances in animal feed" in 2009. This document identified, other than methaemoglobinemy, two toxicological endpoints related to nitrite toxicity: hypertrophy of the adrenal zona glomerulosa in rats, and equivocal evidence for carcinogenesis in female mice.

Moreover, a particular susceptibility of pigs and ruminants (the major food producing animals) to excessive nitrite exposure was reported. This is due to low nitrite reductase activity and to high levels of rumen conversion of exogenous nitrate to nitrite. However, the Panel concluded that the estimated nitrite intakes from feed (for pigs and cattle) were about 5-10 times lower than the respective NOAEL (3.3 mg/kg b.w. per day). So, the Panel reported that concerns for animal health do not subsist under good agricultural practices [23].

Nitrite has been included in the Directive (EC) No 2002/32/EC on undesirable substances in animal feed. In this Directive, maximum limits (expressed as sodium nitrite) in fish meal and complete feeding stuffs (excluding feeding stuffs intended for pets except birds and aquarium fish) equal to 60 and 15 mg kg^{-1}, respectively, were established [24].

Nitrite/Nitrate Accumulation in Different Foodstuffs

Vegetables

The great part of nitrate daily intake derives from vegetable consumption, due to high accumulation capacity of this type of foodstuff [25, 26]. This accumulation takes place especially in the leaves; consequently, leafy vegetables, such as spinach, cabbage and lettuce, may contain high nitrate concentrations [27].

Many factors may influence the nitrate accumulation in vegetables. The most important are timing of N-fertilizers application, water relations, temperature, light exposure, carbon dioxide concentration, soil type and use of herbicides [28, 29].

Nitrites are present at low concentrations in vegetables in which they may derive from the natural conversion of endogenous nitrate [5].

Table 1. Nitrites and nitrates admissible limits in foodstuffs

	Food category	Legal limit
Nitrites	Non-heat-treated processed meat	150 [a]
	Heat-treated processed meat	150 [a]
	Heat-treated processed meat	100 [a]
	Traditionally cured meat products with specific provisions concerning nitrites and nitrates	50-180 [a]
Nitrates	Fresh spinach (*Spinacia oleracea*)	3.500 [b]
	Preserved, deep-frozen or frozen spinach	2.000 [b]
	Fresh Lettuce (*Lactuca sativa L.*) (protected and open-grown lettuce) Harvested 1 October to 31 March:	5.000 [b] (Lettuce grown under cover) 4.000 [b] (Lettuce grown in the open air)
	Fresh Lettuce (*Lactuca sativa L.*) (protected and open-grown lettuce) Harvested 1 April to 30 September:	4.000 [b] (Lettuce grown under cover) 3.000 [b] (Lettuce grown in the open air)
	Iceberg-type lettuce	2.500 [b] (Lettuce grown under cover) 2.000 [b] (Lettuce grown in the open air)
	Rucola (*Eruca sativa, Diplotaxis* sp., *Brassica tenuifolia, Sisymbrium tenuifolium*)	7.000 [b] (Harvested 1 October to 31 March) 6.000 [b] (Harvested 1 April to 30 September)
	Processed cereal-based foods and baby foods for infants and young children	200 [b]
	Ripened cheese	150 [a]
	Whey cheese	150 [a]
	Cheese products	150 [a]
	Dairy analogues, including beverage whiteners	150 [a]
	Non-heat-treated processed meat	150 [a]
	Traditionally cured meat products with specific provisions concerning nitrites and nitrates	10-300 [a]
	Processed fish and fishery products including molluscs and crustaceans	500 [a]

[a] Expressed as mg kg^{-1} of $NaNO_2$ and $NaNO_3$ (Regulation 1129/2011/EC).
[b] Expressed as mg kg^{-1} of NO_3^- (Regulation 1258/2011/EC).

In table 1 the nitrate and nitrite maximum levels admitted as residues by the European Commission Regulation (EC) No. 1258/2011 [30] are reported, while in Figure 1 a graphic reporting the mean nitrate/nitrite levels in vegetables, as indicated in a very useful study recently published [31], is shown.

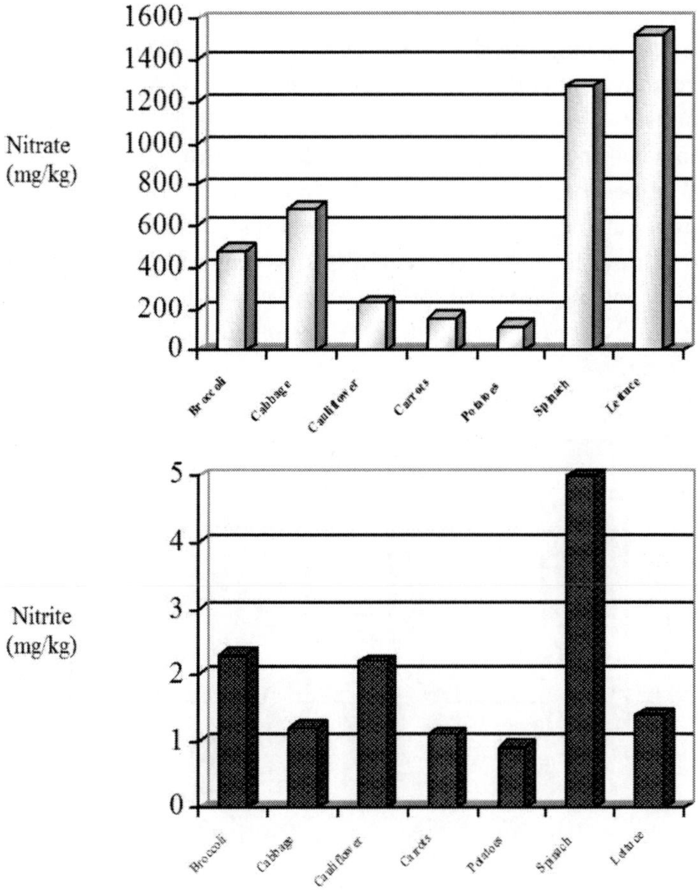

Figure 1. Nitrite and nitrate mean levels in vegetables [13, 29, 31 – 39].

Meats

The relation between nitrite/nitrate and meat products is well-known due to the use of these substances as food additives in meat curing. The wide use of these food additives is due to their capacity to contemporarily exercise many functions (anaerobic microorganisms growth inhibition (in particular of

Clostridium botulinum), fresh meats red color stabilization, product flavour characterization and oxidative rancidity retarding) [40 – 45]. Currently, others food additives able to exercise all these functions are not available; however, in the last years, several studies are deepening alternative stabilization technologies that may replace the use of nitrites and nitrates in different foodstuffs [46, 47].

In some cases, the residual amounts of nitrite and nitrate in cured meats are not corresponding to added dose. This is due to nitrate-nitrite and nitrite-nitrate interconversions that may occur in meat products as consequence of heat treatments and/or lengthened storage (nitrite to nitrate) or of microbiological activity (nitrate reduction to nitrite by nitrate-reducing bacteria (*Micrococcaceae*)) [48 - 51]

Different authors have reported that nitrate is a compound which may be found in fresh meats, at low concentrations, as natural constituent. The same assumption cannot be sustained for nitrite ion [52 – 54].

In Figure 2 a graphic reporting the mean nitrate/nitrite levels in fresh meats is shown.

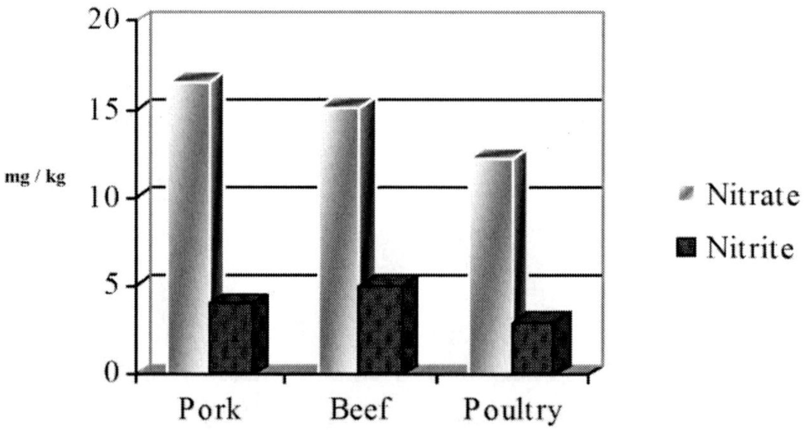

Figure 2. Nitrite and nitrate mean levels in fresh meats [13, 29, 31, 34, 37, 52].

Dairy Products

When water, used in agriculture for irrigation, contains a high concentration of nitrate, as consequence of pollution by fertilizers and/or of different types of wastes and effluents, this compound may be naturally present in raw milk. In cheeses made without the addition of nitrate, its residual level is also directly in concordance with quality of feeding [55].

Other ways of milk pollution by nitrates may be the water used to wash udders, vessels and installations and the negligence in the operation of cleaning after milk removal (when nitric acid is used) [56].

Different authors have studied the natural levels of nitrates and nitrites in dairy products [5, 13, 31, 32, 34, 37, 39, 55, 57 - 59]; a graphical elaboration of these data is shown in Figure 3.

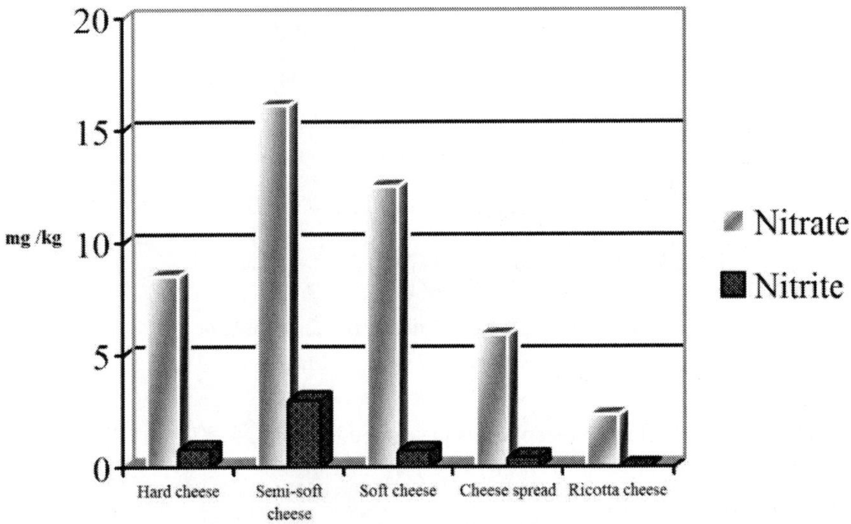

Figure 3. Nitrite and nitrate mean levels in dairy products [13, 31, 32, 34, 37, 39, 55, 57, 58].

Seafood

The relation between nitrate levels in the environment and aquatic productions was studied by different authors especially regarding shellfish, due to the potential of these products to concentrate all chemical and biological contaminants present in water [60, 61]. Nitrate may be used as indicator of biotope conditions [62]; moreover, it may influence the production of algal biotoxins. In particular, Hu et al. reported an increase of cellular toxin (Saxitoxin) and cell density in *Alexandrium tamarense*, up to 76%, when supplemented with $NaNO_3$ [63]; while Selander et al. demonstrated that two strains of *Alexandrium minutum* showed increased cell-specific paralytic shellfish toxin (PST) content after nitrate-rich treatments [64].

For nitrate, there are no maximum levels (or suggested levels) established for shellfish; considering that this compound is involved in many biogeochemical cycles, this may be considered a gap of the actual Legislation. For fish and fish products some limits (i.e. 15 mg kg^{-1}) were established [65].

In Figure 4 a graphic reporting the mean nitrate/nitrite levels in seafood is shown.

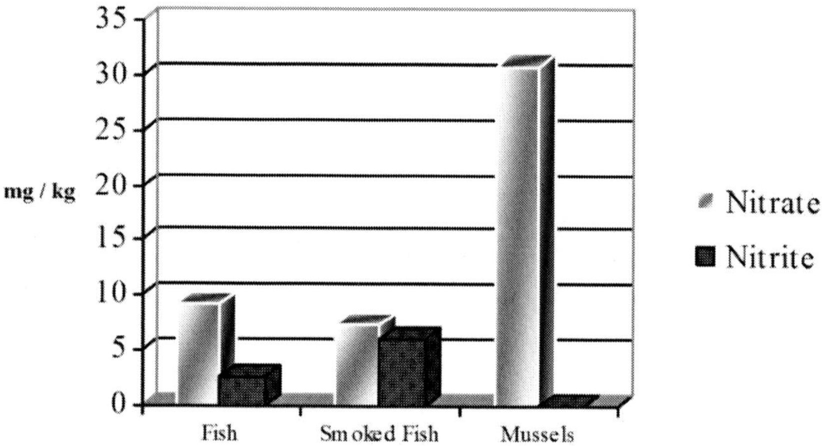

Figure 4. Nitrite and nitrate mean levels in seafood [31, 34, 35, 37, 66, 67].

Experimental

Sampling

The investigations were performed during the period January 2007 – December 2013. The following types of samples were analysed: 150 leafy vegetables (75 spinach, 75 lettuce); 200 fresh meats (50 beef, 50 pork, 50 equine, 50 chicken); 180 cheese samples (107 unripened cheese; 43 ripened cheese; 30 Mozzarella cheese); 150 shellfish (110 mussels, 40 clams); 15 animal feeds (6 feeds for aquaculture, 3 fodder for dairy cows, 2 feeds for pets, 2 feeds for cattle, 2 feeds for veal) (Fig. 5). The samples were collected from local stores located in two Italian regions, Puglia and Basilicata and during official controls.

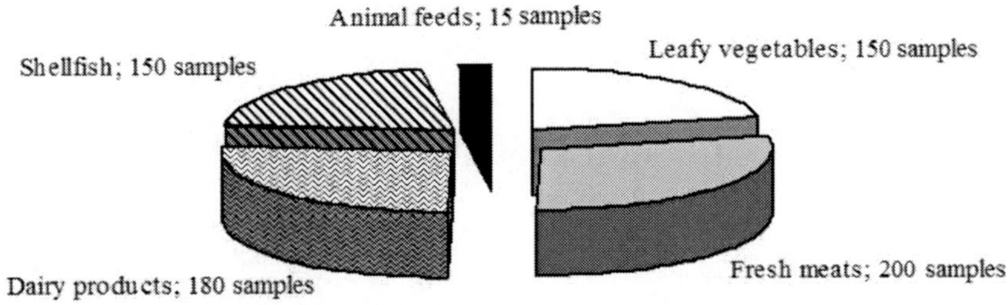

Figure 5. Types of samples analysed.

Analytical Technique

Several methods are currently available for the determination on nitrite and nitrate in different foodstuffs and in animal feeds [68-75]. Among these, ion chromatography is internationally recognized as the most efficient and reliable technique due to its accuracy, specificity and robustness.

The surveys described in this chapter were carried out by using an analytical method that adopts the EN 12014-4:2005 Official Method for the determination of nitrate and nitrite content of meat products [76] together with a chromatographic separation proposed by Dionex Corporation for the determination of inorganic anions [77]. In order to confirm the method reliability and its suitability for purpose, a full validation of this method was developed by following an *in-house* validation model according to the specific European Directives [78, 79] and International Guidelines [80-82]. The most important validation parameters obtained are reported in Table 2 [67].

Sample Preparation

Due to different chemical composition of several types of analysed matrices, it was not possible to use the same procedure of sample extraction and purification; consequently, three different procedures were adopted and validated. The first procedure was used for fresh meats and shellfish samples: 5g of sample, homogenized by mincing machine (fresh meats) or blade homogenizer (shellfish), were mixed with 100 mL of ultrapure water and placed at 70°C for five minutes. The mixture was then cooled and centrifuged at 1500g for 5 min (room temperature). The supernatant was filtered through filter Whatman No. 40 (Whatman, Springfield Mill, UK) and 2 mL of filtrate were filtered through Anotop 10 LC, 0.2 μm, 10 mm filter (Whatman, Springfield Mill, UK) prior to chromatographic analysis.

Table 2. Validation parameters of ion chromatography method

PARAMETER	*NITRITES*	*NITRATES*
Linearity (r)	0.9999	0.9998
Limit of Detection (mg · kg^{-1} in the sample)	1.5	3.2
Limit of Quantification (mg · kg^{-1} in the sample)	4.5	9.6
Specificity (No interferences)	Leafy vegetables Fresh meats Dairy products Shellfish Animal feeds	Leafy vegetables Fresh meats Dairy products Shellfish Animal feeds
CV% (n=6)	4.5 mg · kg^{-1} (LOQ) = 8.5 75 mg · kg^{-1} = 2.6 150 mg kg^{-1} = 3.0 225 mg · kg^{-1} = 2.4	9.6 mg kg^{-1} (LOQ) = 6.7 125 mg kg^{-1} = 3.6 250 mg kg^{-1} = 2.6 375 mg kg^{-1} = 3.9
Mean Recovery	98.7	98.3
Ruggedness (Major Changes: Matrix)	Leafy vegetables Fresh meats Dairy products Shellfish Animal feeds	Leafy vegetables Fresh meats Dairy products Shellfish Animal feeds
Measurement Uncertainty (%)	7.8	9.7

The second procedure was used for dairy products: 4g of sample, homogenized by blade homogenizer, were placed into a 50 mL polypropylene tube and mixed with 40 mL of ultrapure water. The mixture was vortexed for one minute and then centrifuged at 1500g for 5 min (room temperature). The supernatant was filtered through Whatman No. 40 filter (Whatman, Springfield Mill, UK) and 2 mL of filtrate were filtered through Anotop 10 LC, 0.2 μm, 10 mm filter (Whatman, Springfield Mill, UK) prior to chromatographic analysis.

The third procedure was used for leafy vegetables: 100g of sample were homogenized by using an opposite mixer (BUCHI B-400, BUCHI Italia s.r.l., Assago, Milan, Italy), then 200 mL of ultrapure water were added to a 4g portion of homogenized sample and the sample was placed at 70°C for 5 min. After cooling, the mixture was filtered through Whatman No. 41, 150 mm filters (Whatman, Springfield Mill, UK) and then 3 mL of filtrate were purified by using a ISOLUTE® Alumina Neutral Cartridge (Biotage AB, Uppsala, Sweden) previously activated with 3 mL of ultrapure water. After elution, the sample was filtered through Anotop 10 LC, 0.2 μm, 10 mm filter (Whatman, Springfield Mill, UK) prior to chromatographic analysis. A dilution with ultrapure water was effected for samples with nitrates concentrations out of measurement range (9.6 - 625.0 mg kg^{-1}).

High Performance Ion Chromatography

All analytical determinations were carried out by using a Dionex DX500 chromatographic system (Dionex Corporation, Sunnyvale, CA, USA). This system was composed of a quaternary gradient pump (model GP50), a Rheodyne injection valve with a 25 ml injection loop (model RH9125, Cotati, CA, USA), an anion self-regenerating suppressor (model ASRS II, 4 mm) set to 50 mA for chemical suppression and an electrochemical detector (model ED40 set to conductivity mode and with an automatic temperature compensation). The mobile phase consisted of an isocratic elution at 1.0 mL min^{-1} of 9mM Na_2CO_3 (total run time: 20 minutes). The reservoir bottles (DX500 2L bottles; Dionex) were pressurized with pure nitrogen to 0.8 MPa. The chromatographic separations were accomplished by using an IonPac® AS9-HC anion-exchange column (250 x 4mm i.d., particle size 9 μm; Dionex Corporation, Sunnyvale, CA, USA) coupled with a guard column AG9-HC (Dionex Corporation, Sunnyvale, CA). The system was interfaced via proprietary network chromatographic software (PeakNet™, Dionex Corporation, Sunnyvale, CA, USA) to a personal computer for instrumentation control, data acquisition and processing. In Figure 6 a chromatogram related to nitrite and nitrate standard solution is shown.

Saccani and Tanzi [83] reported that some interference compounds of nitrate may be present in some foodstuffs. For this reason, in order to improve the reliability of results that were considered more interesting (high nitrate concentrations and/or quantifiable nitrite concentrations), some samples were also analysed by using a different chromatographic separation. The same HPIC system and sample preparation procedures described above were used for these analytical determinations. The alternative chromatographic column was an IonPac® AS11-HC anion-exchange column (250 mm x 4 mm i.d., particle size: 13 μm; Dionex Corporation), eluted by using a gradient of two solutions: 2 mM NaOH (A) and 21 mM NaOH (B) at a flow rate of 0.5 mL min^{-1}. The gradient program was the following: 100% A isocratic for 13 min, a gradient to 100% B in 1 min, 6 min isocratic and then a re-equilibration for 10 min at 100% A (total run time: 30 minutes). The results obtained from this "confirmatory" technique substantially confirmed those registered by the validated one. Indeed, they were included in the range: concentration measured by validated method ± measurement uncertainty. In Figure 7 two chromatograms related to these "confirmation" analyses, other than to a standard solution, are shown.

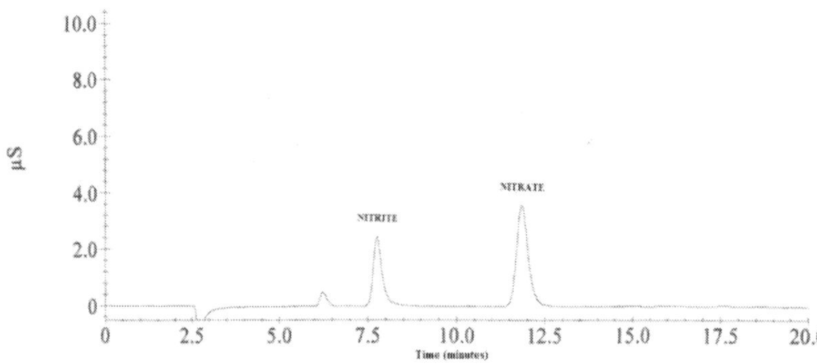

Figure 6. Nitrite and nitrate standard solution (6.5 and 12.5 mg L^{-1} respectively).

Table 3. Results of monitoring

Food category	Number of samples analysed	Number of samples with Nitrites concentration > LOQ	Nitrites concentration range (mean concentration[a])	Number of samples with Nitrates concentration > LOQ	Nitrates concentration range (mean concentration[a])
Leafy vegetables	150	16 (10.7%)	9.5 – 197.5 (46.7)	150 (100%)	15.2 – 5101.0 (945.6)
Fresh meats	200	0 (0%)	-	19 (9.5%)	10.2 – 36.5 (14.7)
Dairy products	180	0 (0%)	-	21 (11.7%)	11.2 – 58.6 (31.5)
Shellfish	150	0 (0%)	-	79 (52.7%)	10.3 – 400.0 (36.9)
Animal feeds	15	0 (0%)	-	8 (53.3%)	11.6 – 187.5 (54.9)

[a] Evaluated only on samples > LOQ.

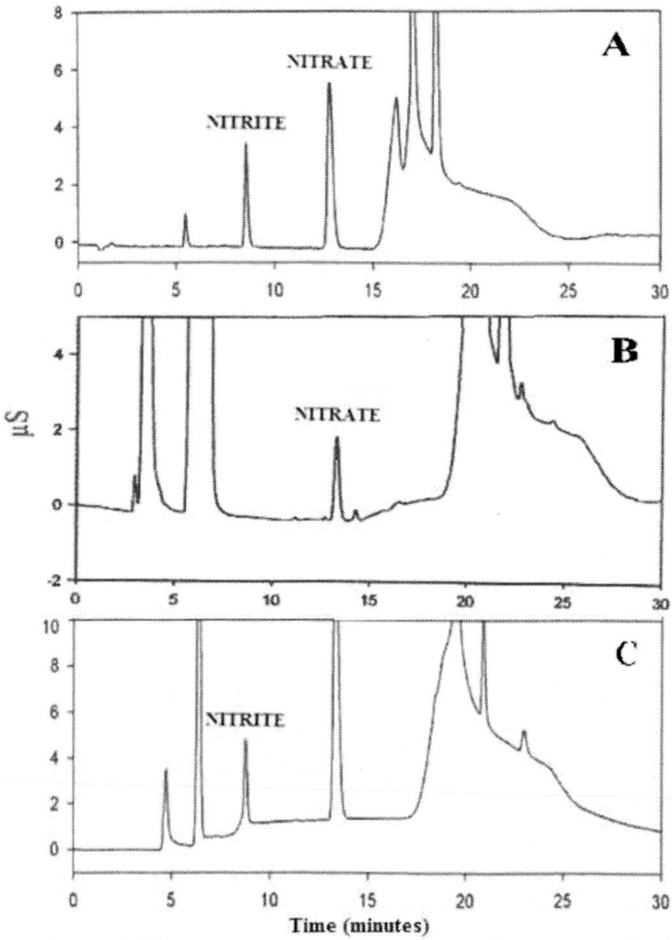

Figure 7. Confirmation analysis: Nitrite and nitrate standard solution (3.0 and 5.0 mg L^{-1} respectively (A); Fresh meat sample (equine) with nitrate content of 36.5 mg kg^{-1} (B); Spinach sample with nitrite content of 163.3 mg kg^{-1} (C).

RESULTS

The results obtained by analysing 680 samples of different foodstuff categories (leafy vegetables, fresh meats, dairy products and shellfish), other than 15 animal feed samples, are reported in Table 3 and schematized in Figure 8.

Leafy Vegetables

As well known, nitrates are present at high levels in leafy vegetables. This was confirmed in this survey, considering that concentrations up to 2978.1±288.9 mg kg^{-1} (a frozen spinach sample) with a mean value equal to 914.7 mg kg^{-1} were obtained for spinach analysis, and concentrations up to 5101.0±494.8 mg kg^{-1} (a Butterhead lettuce sample) with a mean value equal to 976.6 mg kg^{-1} were obtained for lettuce analysis. It is also important to underline that one sample of spinach of IV range, three samples of frozen spinach, one sample of Butterhead lettuce, of Trocadero lettuce and of Scarola lettuce and two samples of Romana lettuce (6.0% of total analysed samples) were characterized by a nitrate concentration higher than the legal limit.

Quantifiable amounts of nitrites were registered. This result may be considered very important considering that nitrites residues are not allowed by the actual Normative. Nitrite quantifiable amounts were detected especially in spinach samples, both at low levels (12 samples with a nitrite level lower than 28.5 mg kg^{-1}) and at high concentrations: one frozen spinach sample (106.8±8.3 mg kg^{-1}) and two fresh spinach (IV range) samples (173.7±13.5 mg kg^{-1} and 197.5±15.4mg kg^{-1}). Only one sample of lettuce (Romana) showed a similar result (66.5±5.2 mg kg^{-1} of nitrite). The authors suppose that particular conditions (i.e. environmental pollution, acid rains, etc.) may bring about a strong reduction of natural nitrate to nitrite. These topics will be explored in greater depth in future studies.

In Figure 9 the chromatograms related to the Butterhead lettuce sample with 5101.0±494.8 mg kg^{-1} of nitrate, and to a fresh spinach (IV range) sample with a high nitrite concentration are shown.

The percentage of samples resulted '*not-compliant*' for nitrates was non-negligible (6.0%), so, these results confirm the importance of official controls. Another important result is related to nitrite levels, since it is possible to affirm that these residues may be present, at non-negligible concentrations, especially in spinach. Therefore, the authors suggest the introduction of a 'maximum admissible level' for nitrites, relating to leafy vegetables, in Communities Regulations [84].

Fresh Meats

Two hundred fresh meat cuts (50 beef, 50 pork, 50 equine and 50 chicken) were collected from local stores, minced in laboratory and then analysed. No nitrite quantifiable amounts were detected in all analysed samples; consequently, it is possible to affirm that this residue cannot be considered as "natural" in fresh meats and then admitted. The same consideration is valid for nitrate, but only relating to chicken fresh meats.

For beef, pork and equine fresh meats, nitrate quantifiable residues may be detected. In particular, concentrations higher than the method limit of quantification (LOQ = 9.6 mg kg^{-1}) were registered in 19 samples (9.5% of total analysed samples). These quantifiable residues were obtained in 8 beef, 6 pork and 5 equine meat samples, at concentrations in the range 10.2 – 36.5 mg kg^{-1} (mean value: 14.7 mg kg^{-1}). By analyzing the obtained data by ANOVA one-way test, it is possible to affirm that the mean concentrations detected in beef and pork fresh meats were similar, while in equine fresh meats the nitrate mean concentration is higher. This consideration is in accordance with the higher concentration which was found by analysing an equine fresh meat sample (36.5 mg kg^{-1}) (beef and pork samples never exceed 15.0 mg kg^{-1}). In order to justify this difference, the authors make reference to equine feeding which is particularly full of nitrogenous compounds [85]. Moreover, the authors suggest the introduction of a maximum admissible limit for nitrate in fresh meats, tentatively estimated in 30.0 mg kg^{-1} for beef and pork fresh meats and in 40.0 mg kg^{-1} for equine fresh meats [86]. In Figure 9 a chromatogram related to an equine fresh meat sample, with a quantifiable nitrate concentration, is shown.

Dairy Products

One hundred eighty cheese samples were analysed, subdivided as follows: unripened cheeses (107 samples), ripened cheeses (43 samples), cow's milk Mozzarella cheeses (15 samples) and buffalo-milk Mozzarella cheeses (15 samples). As reported for fresh meats, also for dairy products it is possible to exclude the nitrite presence at quantifiable amounts. Consequently, the 'natural' presence of this compound in dairy products was not demonstrated. The same result was registered for nitrate in cow's milk Mozzarella cheeses. Speaking of which, it is interesting to underline the evident difference between results related to these samples and those related to buffalo-milk Mozzarella cheese. Indeed, 6 buffalo-milk Mozzarella cheese samples on 15 (40% of total samples analysed) showed a quantifiable amount of nitrate. This result suggests a new topic to deepen in the future.

Quantifiable amounts of nitrate were detected in 21 samples (12.7% of total analysed samples) as subdivided: 10 unripened cheese samples, 5 ripened cheese samples and 6 buffalo-milk Mozzarella cheese samples. The higher concentration was registered for an unripened cheese sample (58.6 mg kg^{-1}).

The nitrate levels measured, the measurement uncertainty of the method (9.7%) and a reasonable tolerance were taken into account for an estimation of a maximum admissible limit relating to nitrates in dairy products. The authors suggest a value equal to 70.0 mg kg^{-1}. In Figure 9 a chromatogram related to a cheese sample, with a quantifiable amount of nitrate, is shown. [87].

Shellfish

No quantifiable residues of nitrite and nitrate were detected in clam samples. This result is largely justified considering the low accumulation potential which characterizes this type of bivalves [88-90]. As conclusion, it is possible to exclude quantifiable amounts of nitrite and nitrate in clams.

The results related to mussel samples are very different and more interesting.

Nitrite residues were not quantified and, consequently, they cannot be admitted as residue in mussels.

Seventy nine samples (71.8% of total analysed samples) showed a quantifiable amount of nitrates. The most part of these samples (73) did not exceed 42.4 mg kg^{-1}. These concentrations were elaborated in order to calculate and suggest a maximum admissible level for nitrates in mussels. Considering the higher frequency registered (in the range 10.0 – 14.1 mg kg^{-1}) increased of 3σ (σ = 8.2 mg kg^{-1}) and the measurement uncertainty of the method (9.7%), a value equal to 49.1 mg kg^{-1} was obtained. This value can be easily rounded off to 50.0 mg kg^{-1} which was the suggested maximum admissible limit.

Six samples showed a nitrate concentration higher than 100 mg kg^{-1}, and precisely: 114.1, 227.3, 225.7, 250.9, 226.3 and 400.0 mg kg^{-1}. Considering that certain strains of *Escherichia coli* may grow better in the presence of high levels of nitrates, these samples were also analysed in terms of microbiological contamination (*Salmonella* and *Escherichia coli*), obtaining a rather correlation between nitrate levels and *Escherichia coli* contamination (in 2 samples the *Escherichia coli* contamination was higher than legal limit).

This result may be considered very important relating to a possible introduction of nitrate as new indicator of mussels safety; the authors named it "mussels safety indicator" (MSI) [91]. In Figure 9 a chromatogram related to a mussels sample characterized by a high nitrate concentration is shown.

Animal Feeds

Fifteen different types of animal feeds were analysed. No nitrite residues were quantified, so, the same consideration already stated for fresh meats, dairy products and shellfish is also valid for animal feeds. As shown in Figure 10, nitrate results were very variable. 7 samples showed not quantifiable amounts of nitrates, while other samples were contaminated by nitrates at levels up to 187.5 mg kg^{-1} observed in a feed for veal. This high variability of obtained results demonstrated the necessity to develop most controls for this type of matrix. Moreover, the authors suggested the introduction of a maximum admissible limit for nitrate in animal feeds, since some countries make reference only to nitrite (that must be absent) in their regulations [92]. In Figure 9 a chromatogram related to feed for veal, characterized by a high nitrate concentration, is shown.

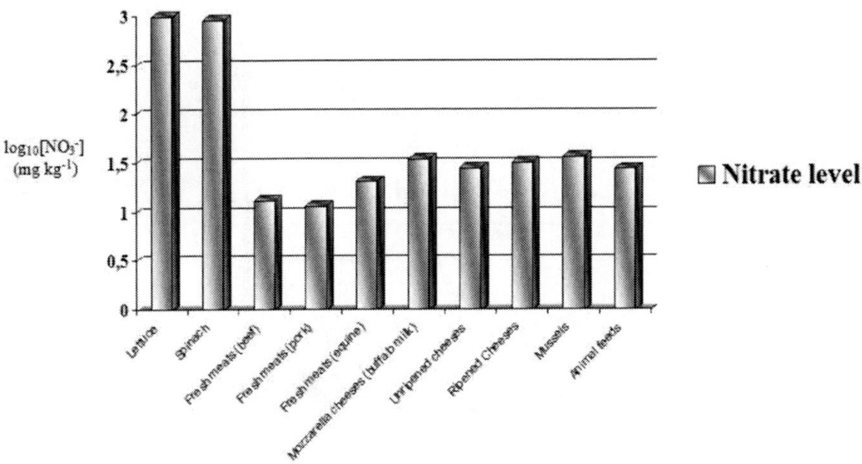

Figure 8. Mean nitrate levels detected in this survey.

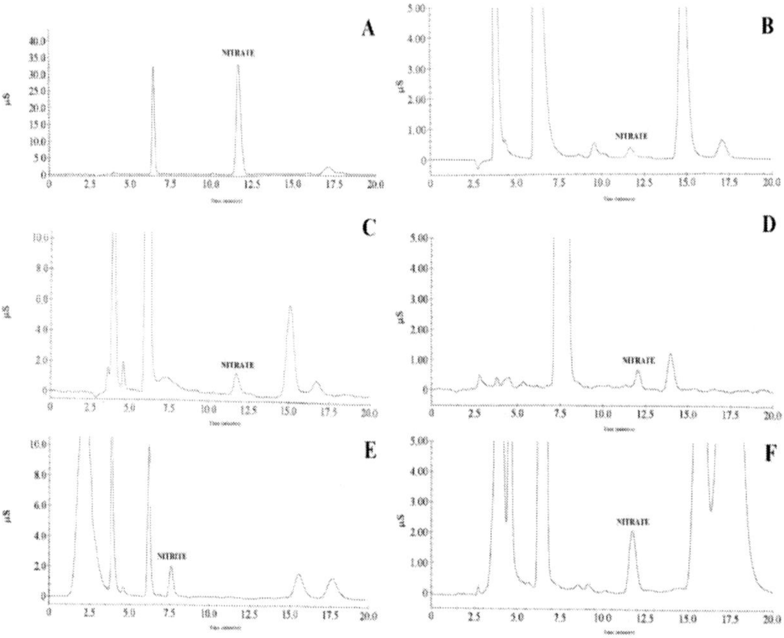

Figure 9. Chromatograms comparison: Butterhead lettuce sample with 5101.0 mg kg^{-1} of nitrate (A); Equine fresh meat sample with 23.5 mg kg^{-1} of nitrate (B); Unripened cheese sample with 58.6 mg kg^{-1} of nitrate (C); Mussel sample with 42.4 mg kg^{-1} of nitrate (D); Fresh spinach (IV range) sample with 173.7 mg kg^{-1} of nitrite (E); Feed for veal with 108.6 mg kg^{-1} of nitrate (F).

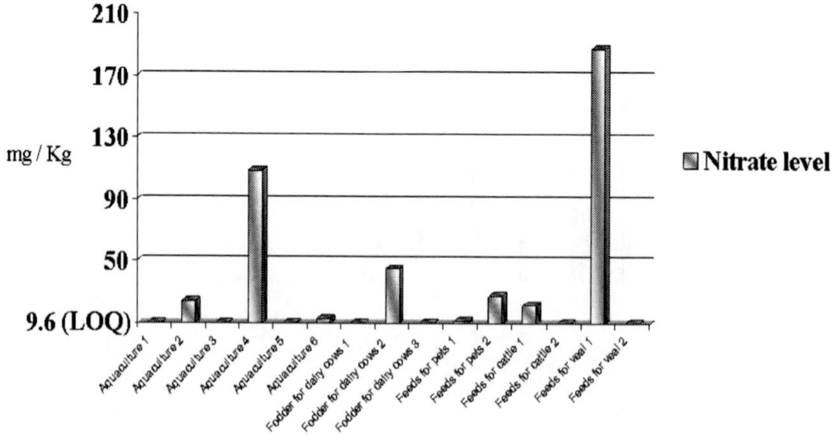

Figure 10. Nitrate levels detected in animal feed samples.

CONCLUSION

The results obtained through several monitoring, carried out in order to identify and quantify the nitrite and nitrate levels in wide consumption foodstuffs (leafy vegetables, fresh meats, dairy products and shellfish) and animal feeds, are reported and described in this chapter.

The analyses were carried out by using a validated ion chromatography with conductivity detection method and the more interesting and significant results were confirmed through an alternative ion chromatography method.

High levels of nitrate were detected, especially in leafy vegetables and in some samples of mussels and animal feeds.

The actual Normative does not foresee a maximum admissible level for nitrate relating to some types of foodstuff (fresh meats, dairy products and mussels). However, several quantifiable amounts of nitrates were detected in these products. Consequently, in order to fill a legislative gap and to allow a correct interpretation of official controls results, the authors suggest the introduction of new legal limits related to nitrate in these matrices.

Four samples of spinach and 5 of lettuce with nitrate concentrations higher than the actual legal limit were registered; moreover, some mussel samples with high nitrate levels (up to 400.0 mg kg^{-1}) must be taken into consideration for future studies.

As it regards nitrite, this contaminant was detected only in leafy vegetables, both at low levels (lower than 28.5 mg kg^{-1}) and at high levels (up to 197.5 mg kg^{-1}), by suggesting the introduction of specific legal limits.

The results reported in this chapter confirm that the problem related to nitrate/nitrite accumulation in foodstuffs cannot be underestimated and they highlight the necessity to develop most controls.

ACKNOWLEDGMENTS

Ministero della Salute (Rome, Italy) is gratefully acknowledged for providing the financial support.

REFERENCES

[1] World Health Organization – International Agency for Research on Cancer. IARC Monographs on the Evaluation of Carcinogenic Risks to Humans, Vol. 94, *Ingested Nitrate and Nitrite, and Cyanobacterial Peptide Toxins.* Lyon, France, 2010.

[2] Fields, S. (2004). Global nitrogen: cycling out of control. *Environ Health Perspect.* 112, A557–A563.

[3] Hill, AR. (1996). Nitrate removal in stream riparian zones. *Journal of Environmental Quality*, 25(4), 743–754.

[4] Walters, CL. Nitrite and nitrate in foods. In: Hill MJ ed. *Nitrates and nitrites in food and water.* Cambridge, UK: Woodhead Publishing Ltd. 1996; 93.

[5] Schuddeboom, L. J. *Nitrates and nitrites in foodstuffs.* Council of Europe. Strasbourg, Belgium. 1993.

[6] European Commission. Reports of the Scientific Committee for Food (thirty-eighth series). *Opinions of the Scientific Committee for Food on: Nitrates and Nitrite.* European Commission Directorate-General Industry. Brussels-Luxembourg. 1997.

[7] Archer, MC. (1989). Mechanisms of action of N-nitroso compounds. *Cancer Surveys*, 8, 241–250.

[8] Sen, NP; Iyengar, JR; Donaldson, BA; Panalaks, T. (1974). Effects of sodium nitrite concentration on the formation of nitrosopyrrolidine and dimethylnitrosamine in fried bacon. *Journal of Agriculture and Food Chemistry*, 22, 540-541.

[9] Krol, B. *Proceedings of the International Symposium on Nitrite in Meat Production: Held at the Central Institute for Nutrition and Food Research TNO,* Zeist, the Netherlands, September 10-14. Pudoc, Wageningen. 1974.

[10] Gray JI. (1976). N-nitrosamines and their precursors in bacon: A review. *J. Milk Food Technol.* 39, 686.

[11] National Food Agency of Denmark. *Report: Investigations on the effects of nitrite in commercially prepared Danish cured meat products.* Søborg, Denmark. 1983.

[12] National Academic of Sciences. *Committee on Nitrite and Alternative Agents in Food.* National Academy Press, Washington DC. 1982.

[13] Thomson, B; Swallow, B; Mitchell, J. *Nitrates and nitrites dietary exposure and risk assessment. New Zealand Food Safety Authority.* Christchurch, New Zeland. 2004.

[14] United States Environment Protection Agency. *Nitrates and Nitrites Teach Chemical Summary* - U.S. EPA, Toxicity and Exposure Assessment for Children's Health. Chicago, USA. 2007.

[15] Agency for Toxic Substances and Disease Registry, Division of Toxicology and Environmental Medicine. *Nitrates and nitrites.* Atlanta, GA, USA. 2011.

[16] European Union. The EU Nitrates Directive. 2010. Bruxelles, Belgium.

[17] European Community. EC Directive relating to the quality of water intended for human consumption. *Official Journal of the European Community* L229, 80/778/EEC. 1980.

[18] United States Environmental Protection Agency – USEPA. Maximum contaminant levels (subpart B of part 141, National primary drinking-water regulations), *US Code of Federal Regulations*, Title 40, parts 100–149. US Environmental Protection Agency, Washington, DC, 1986, 524–528.

[19] Böhlke, JK. (2002). Groundwater recharge and agricultural contamination. *Hydrogeology Journal*, 10, 438–439.

[20] World Health Organization. *Nitrate and Nitrite in Drinking-water - Background document for development of WHO Guidelines for Drinking-water Quality.* Geneva, Switzerland. 2011.

[21] Faulkner, DB; Hutjens, MF. (1989). Nitrates in livestock feed. American Society of Animal Science Symposium on nitrate toxicity. Available at http://www.livestocktrail.illinois.edu/uploads/beefnet /papers/nitrates.pdf

[22] Yaremcio B. (1991). Nitrate Poisoning and Feeding Nitrate Feeds to Livestock. *Agri-Facts*, Agdex 400/60-1, 1-5.

[23] European Food Safety Authority. (2009). Nitrite as undesirable substances in animal feed - Scientific Opinion of the Panel on Contaminants in the Food Chain. *The EFSA Journal*, 1017, 1-47.

[24] European Commission. Directive 2002/32/EC of the European Parliament and of the Council of 7 May 2002 on undesirable substances in animal feed. *Official Journal of the European Union*, 30 May 2002, L140/10-21.

[25] European Food Safety Authority. (2009). Nitrate in vegetables - Scientific Opinion of the Panel on Contaminants in the Food Chain. *The EFSA Journal*, 689, 1–79.

[26] Corrè, WJ; Breimer, T. (1979). *Nitrate and nitrite in vegetables. Literature Survey,* Vol. 39. Center for Agricultural Publishing Documentation, Wageningen, Netherlands.

[27] Brian, NS; Loscalzo, J. History of nitrite and nitrate in food. In: Brian, NS; Loscalzo, J Eds. *Nitrite and Nitrate in Human Health and Disease.* Humana Press, Springer, New York. 2011.

[28] Coördinatie-Commissie voor de metingen van Radioactiviteit en Xenobiotische Stoffen. (1988). *Nitrogen and nitrogen compounds in environment and nutrition in The Netherlands.* CCRX, Ministry of Housing, Physical Planning and Environment Control. The Hague, Netherlands.

[29] National Academy of Sciences. The health effects of nitrate, nitrite and *N*-nitroso compounds: In: *Part 1 of a 2-Part Study by the Committee on Nitrite and Alternative Curing Agents in Foods.* National Academy Press, Washington, DC. 1981.

[30] European Commission. Commission Regulation (EU) No 1258/2011 of 02 December 2011 amending Regulation (EC) No 1881/2006 as regards maximum levels for nitrates in foodstuffs. *Official Journal of the European Union*, 03 December 2011, L320, 15–17.

[31] Griesenbeck, JS; Steck, MD; Huber Jr, JC; Sharkey, JR; Rene, AA; and Brender, JD. (2009). Development of estimates of dietary nitrates, nitrites, and nitrosamines for use with the short willet food frequency questionnaire. *Nutr J.*, 8, 16.

[32] French National Inventory: Ministeres de I'Environment, *de I'Agriculture, de la Rechereche et de la Technologie, de I'lndustrue et de la Sante.* Paris. 1982.

[33] Siciliano, J; Krulick, S; Heisler, EG; Schwartz, JH; White, JW. (1975). Nitrate and nitrite content of some fresh and processed market vegetables. *J. Agric. Food Chem.*, 23, 461–464.

[34] Howe, GR; Harrison, L; Jain, M. (1986). A short diet history for assessing dietary exposure to *N*-nitrosamines in epidemiologic studies. *Am. J. Epidemiol.*, 124, 595–602.

[35] Food Standards Agency: Joint Food Safety Standards Group. *Food Surveillance Information Sheet.* Ministry of Agriculture, Fisheries, and Food. London, UK. 1998.

[36] Petersen, A; Stoltze, S. (1999). Nitrate and nitrite in vegetables on the Danish market: content and intake. *Food Additives and Contaminants*, 16 (7), 291-299.

[37] Pobel, D; Riboli, E; Cornée, J; Hémon, B; Guyader, M. (1995). Nitrosamine, nitrate and nitrite in relation to gastric cancer: a case-control study in Marseille, France. *Eur. J. Epidemiol.*, 11(1), 67–73.

[38] Merino, L; Darnerud, PO; Edberg, U; Aman, P; Castillo, MDP. (2006). Levels of nitrate in Swedish lettuce and spinach over the past 10 years. *Food Additives and Contaminants*, 23, 1283–1289.

[39] VonCollett, P. (1983). Beitrag zur bestimmung von Nitrit und Nitrat in lebensmitteln. *Deut Lebensm-Rundsch*, 79, 370–375.

[40] Iammarino, M; Di Taranto, A. Evaluation of natural levels of substances commonly-used as food additives in fresh meat preparations. In: Ortega MP, Soto R editors. *Meat consumption and health.* NY, USA: Nova Publisher, Inc; 2013; 153-170.

[41] Pierson, M. D. & Smoot, L. A. (1982). Nitrite, nitrite alternatives and the control of Clostridium botulinum in cured meats. *Critical Reviews in Food Science and Nutrition*, 17, 141-187.

[42] Noel, P; Briand, E; Dumont, JP. (1990). Role of nitrite in flavour development in uncooked cured meat products: sensory assessment. *Meat Science*, 28, 1-8.

[43] Chichester, CO; Mrak, EM. Cured meat flavour. In: Chichester CO, Mrak EM, Schweigert BS editors. *Advances in food research.* Orlando, Florida, USA: Academic Press Inc; 1984; 10-11.

[44] Kanner, J. (1994). Oxidative processes in meat and meat products: quality implication. *Meat Science*, 36, 673-676.

[45] Skibsted, LH. Cured meat products and their oxidative stability. In: Johnston DE, Knight MK, Ledward DA editors. *The chemistry of muscle-based foods*. Cambridge, UK: The Royal Society of Chemistry; 1992; 266-286.

[46] Sebranek, JG; Bacus, JN. (2007). Cured meat products without direct addition of nitrate or nitrite: what are the issues?. *Meat Science*, 77, 136 – 147.

[47] Stringer, SC; Pin, C. *Microbial risks associated with salt reduction in certain foods and alternative options for preservation.* Food Safety and Computational Microbiology Group, IFR enterprises. Institute of Food Research. Norwich, UK. 2005.

[48] Toldrà, F. *Handbook of meat processing*. John Wiley & Sons edition. Malden, USA: Blackwell Publishing; 2010.

[49] Martìn, A; Colìn, B; Aranda, E; Benito, MJ; Còrdoba, MG. (2007). Characterization of Micrococcaceae isolated from Iberian dry-cured sausages. *Meat Science*, 75(4), 696-708.

[50] Honikel, KO. (2007). The use and control of nitrate and nitrite for the processing of meat products. *Meat Science*, 78, 68-76.

[51] Hui, YH, Nip, WK. Micrococcaceae. In: Hui YH, editor. *Food Biochemistry and Food Processing.* New York, NY, USA: Wiley-Blackwell Publishing; 2006; 644-645.

[52] Tanzi, E; Saccani, G. (2005). Nitrati in carni e prodotti stagionati. *Laboratorio 2000*, 3, 44-47.

[53] Bernini, R; Picconi, F; Iacovacci, V; Abballe, F. (2001). Analisi di salumi mediante cromatografia ionica - II: Nitrito e Nitrato. *Industrie Alimentari*, XL, 741-746.

[54] Hsu, J., Jayashree, A. & Lee, N.A. (2009). Nitrate and nitrite quantification from cured meat and vegetables and their estimated dietary intake in Australians. *Food Chemistry*, 115, 334–339.

[55] Tudor, L; Mitrânescu, E; Tudor, L; Furnaris, F. *Assessment of nitrate and nitrite content of Romanian traditional cheese*. Lucrâri Stiintifice Medicinâ Veterinariâ, Timisoara, XL. 2007.

[56] Marius, O; Vasile, B. (2010). Research on the level of nitrate in milk and some dairy products. Analele Universităţii din Oradea Fascicula: Ecotoxicologie, Zootehnie şi Tehnologii de Industrie Alimentară. Available at: http://protmed.uoradea.ro/facultate/anale/ecotox_zooteh_ind_alim/2010/ipa/108%20Osvat%20Marius.pdf.

[57] Iammarino, M; Di Taranto, A. (2013). Development and Validation of an Ion Chromatography Method for the Simultaneous Determination of Seven Food Additives in Cheeses. *Journal of Analytical Sciences, Methods and Instrumentation*, 3, 30-37.

[58] Topçu, A.; Topçu, AA; Saldamli, I; Yurttagül, M. (2006). Determination of Nitrate and Nitrite Content of Turkish Cheeses. *African Journal of Biotechnology*, 5(15), 1411-1414.

[59] Zamrik, MA. (2013). Determination of Nitrate and Nitrite Contents of Syrian White Cheese. *Pharmacology & Pharmacy*, 4(2), 171-175.

[60] Pruell, RJ; Quinn, JG; Lake, JL; Davis, WR. Availability of PCBs and PAHs to *Mytilus edulis* from artificially resuspended sediments. In: Capuzzo JM and Kester DR editors. Oceanic processes in marine pollution, vol. 1: *Biological processes and wastes in the ocean*. Malabar, FL: Krieger; 1987; 97-108.

[61] Wang, WX; Fisher, NS; Luoma, SN. (1996). Kinetic determinations of trace element bioaccumulation in the mussel *Mytilus edulis. Marine Ecology Program Series*, 140, 91-113.

[62] Douda, K. (2010). Effects of nitrate nitrogen pollution on Central European unionid bivalves revealed by distributional data and acute toxicity testing. Aquatic *Conserv: Mar. Freshw. Ecosyst.*, 20, 189–197.

[63] Hu, H; Chen, W; Shi, Y. Cong, W. (2006). Nitrate and phosphate supplementation to increase toxin production by the marine dinoflagellate *Alexandrium tamarense. Marine Pollution Bulletin*, 52, 756–760.

[64] Selander, E; Cervin, G; Pavia, H. (2008). Effects of nitrate and phosphate on grazer-induced toxin production in *Alexandrium minutum. Limnol. Oceanogr.*, 53(2), 523–530.

[65] Canadian Food Inspection Agency. Canadian guidelines for chemical contaminants and toxins in fish and fish products. Fish products standards and methods manual. Ottawa, Ontario. 2011.

[66] Cornee, J; Lairon, D; Velema, J; Guyader, M; Berthezene, P. (1992). An estimate of nitrate, nitrite, and N-nitrosodimethylamine concentrations in French food products or food groups. *Sciences des Aliments*, 12,155-197.

[67] Iammarino, M; Di Taranto, A; Cristino, M. (2012). Endogenous levels of nitrites and nitrates in wide consumption foodstuffs: Results of five years of official controls and monitoring. *Food Chemistry*, 140, 763-771.

[68] Association of Analytical Chemists. AOAC Official Method 935.48. Nitrates and Nitrites in Meat, Xylenol Method. 39.1.20. *Official Methods of Analysis of AOAC International,* 18th Ed., Gaithersburg, MD. p8. 2000.

[69] Association of Analytical Chemists. AOAC Official Method 976.14. Nitrate and Nitrite in Cheese, Modified Jones Reduction Method. 33.7.16. *Official Methods of Analysis of AOAC International,* 18th Ed. Gaithersburg, MD. p73. 2000.

[70] Sjöberg, AM; Alanko, TA. (1994). Spectrophotometric determination of nitrate in baby foods: collaborative study. *J. AOAC Int.*, 77, 425-30.

[71] Association of Analytical Chemists. AOAC Official Method 973.31. Nitrites in Cured Meat, Colorimetric Method. Codex-Adopted-AOAC Method. 39.1.21. *Official Methods of Analysis of AOAC International,* 18th Ed. Gaithersburg, MD. p8. 2000.

[72] Association of Analytical Chemists AOAC Official Method 951.03 Nitrogen/Nitrite in Grains/Flour. *Official Methods of Analysis of AOAC International,* 18th Ed. Gaithersburg, MD. 2000.

[73] Association of Analytical Chemists AOAC Official Method 968.07 Nitrogen (Nitrates and Nitrites) in Animal Feed. *Official Methods of Analysis of AOAC International,* 18th Ed., Gaithersburg, MD. 2000.

[74] Zhong, I; Chen, P; Yu, X; Xu, D; Fang, K; Peng, J. (2012). Determination of nitrite and nitrate in dairy products by improved ion chromatography. *Se. Pu.*, 30(6), 635-40.
[75] Siu, DC; Henshall, A. (1998). Ion chromatographic determination of nitrate and nitrite in meat products. *J. Chromatography A*, 804(1-2), 157-60.
[76] European Normalization Committee. (2005). Standard Reference EN 12014–4:2005. Foodstuffs – Determination of nitrate and/or nitrite content – Part 4: Ion exchange chromatographic (IC) method for the determination of nitrate and nitrite content of meat products. CEN-CENELEC Management Centre, Brussels.
[77] Dionex Corporation. (2007). Product manual for IonPac AS9-HC. *Dionex Reference Library version 31*, 14.
[78] European Commission. Regulation (EC) No 882/2004 of the European Parliament and of the Council of 29 April 2004 on official controls performed to ensure the verification of compliance with feed and food law, animal health and animal welfare rules. *Official Journal of the European Union*, 30 April 2004, L165, 1–141.
[79] European Commission. Commission Decision 2002/657/EC of 12 August 2002 implementing Council Directive 96/23/EC concerning the performance of analytical methods and the interpretation of results. *Official Journal of the European Union*, 17 August 2002, L221, 8–36.
[80] Miller, EJC; Miller, JN. *Statistics for Analytical Chemistry,* 3rd ed. New York, USA: Ellis Horwood PTR Prentice Hall; 1993.
[81] Youden, WJ; Steiner, EH. *Statistical Manual of the AOAC.* Association of the Official Analytical Chemists. Washington DC; 1975.
[82] Thompson, M; Ellison, SLR; Wood, R. (2002). Harmonized guidelines for single laboratory validation of methods of analysis. *Pure Appl. Chem.*, 74 (5), 835-855.
[83] Saccani, G; Tanzi, E. (2006). Determination of Nitrite, Nitrate, and Glucose-6-Phosphate in Muscle Tissues and Cured Meat by IC / MS. *Journal of AOAC International*, 89, 712–719.
[84] Iammarino, M; Di Taranto, A; Cristino, M. (2013). Monitoring of nitrites and nitrates levels in leafy vegetables (spinach and lettuce): a contribution to risk assessment. *J. Sci. Food Agric.*, 94, 773–778.
[85] National Research Council. *Nutrient Requirements of Equines,* 5th ed. Washington DC: National Research Council. 1989. Chapter 4.
[86] Iammarino, M; Di Taranto, A. (2012). Nitrite and nitrate in fresh meats: a contribution to the estimation of admissible maximum limits to

introduce in directive 95/2/EC. *International Journal of Food Science and Technology*, 47, 1852–1858.

[87] Iammarino, M; Di Taranto, A; Palermo, C; Muscarella, M. Assessment of Natural Levels of Substances with Preservative Effects in Dairy Products. In: Castelli H, du Vale L editors. *Handbook on Cheese: Production, Chemistry and Sensory Properties.* NY, USA: Nova Publisher, Inc; 2013; 559-571.

[88] Chalkiadakis, O; Paraskevopoulou, V; Roussos, I; Dassenakis, M; Lydakis Simantiris, N. (2013). Comparative Study of the Accumulation of Ni in Different Tissues of Mussels and Soft Clams. *E3S Web of Conferences*, 1, 11003.

[89] Fang, ZQ; Cheung, RYH; Wong, MH. (2003). Heavy metals in oysters, mussels and clams collected from coastal sites along the Pearl River Delta, South China. *Journal of Environmental Sciences*, 15(1), 9 – 24.

[90] Hess, P; Abadie, E; Hervé, F; Berteaux, T; Séchet, V; Aráoz, R; Molgó, J; Zakarian, A; Sibat, M; Rundberget, T; Miles, CO; Amzil, Z. (2013). Pinnatoxin G is responsible for atypical toxicity in mussels (*Mytilus galloprovincialis*) and clams (*Venerupis decussata*) from Ingril, a French Mediterranean lagoon. *Toxicon*, 75, 16-26.

[91] Iammarino, M; Di Taranto, A; La Salandra G. Correlation between nitrate levels and microbiological safety in shellfish (mussels and clams): a preliminary study. In: .... editors. Shellfish: Human Consumption, Health Implications and Conservation Concerns. NY, USA: Nova Publisher, Inc. 2014. *in press.*

[92] Italian Ministry of Health. PNAA 2012-2013-2014. Addendum 1/2013. Piano Nazionale di controllo ufficiale sull'alimentazione degli animali 2012-2013-2014. Rome, Italy. 2013.

In: Nitrogen Fertilizer
Editor: Karlene Winslow
ISBN: 978-1-63321-585-6

Chapter 2

PRODUCTION OF GRANULAR UREA AS NITROGENOUS FERTILIZER

Ivana M. Cotabarren, Diego E. Bertin, Sussy Veliz Moraga, Leonardo Mirazú, Juliana Piña and Verónica Bucalá**

Department of Chemical Engineering, PLAPIQUI,
Universidad Nacional del Sur, CONICET, Bahía Blanca, Argentina

ABSTRACT

The demand for urea is continuously growing and entwined with the need for fertilizers and animal feed additives.

Industrially, urea is initially produced in liquid form as a concentrated solution. Then, it can be converted into particulate material either through granulation or prilling processes. Since granules have better attributes than prills, nowadays granulation is the preferred production route. Urea granulation is a multifaceted process that requires several operation units, which constitute the granulation circuit, to produce the solid form (granules) with the desired attributes. The main unit of the circuit is the granulator, where small urea particles known as seeds are continuously fed and sprayed with a urea concentrated solution. The seeds grow through deposition of the solution droplets onto the solids surface, followed by water evaporation and urea solidification. The

* To whom correspondence should be addressed. Tel: 54-291-486-1700, Fax: 54-291-486-1600, E-mail: vbucala@plapiqui.edu.ar.

granules that leave the size enlargement unit are size classified into product, oversize and undersize streams. The product is transported to storage facilities, while the oversize fraction is fed to crushers for size reduction. The crushed oversize particles are then combined with the undersize granules and recycled back to the granulator as seeds.

Focusing on urea, the advantages of granules over prills are discussed by exploring the physical properties of both solid forms. Then, the current available technologies for urea granulation are presented in a comparative manner. From this analysis, the fluidized-bed granulator appears as the most widely used equipment for granular urea production. Due to this preference, different approaches to model fluidized-bed granulators are presented aiming to give a comprehensive picture of the fundamental phenomena that occurs within these granulation units. Special attention is placed on the granules growth mechanism, and its proper representation. Although coating is the preferred urea growth mechanism, unexpected operating situations may favor size enlargement by agglomeration that is an undesired phenomenon. Therefore, based on experimental data obtained in a pilot-scale fluidized-bed batch granulator for urea production, the influence of the operating variables on both granules quality and growth mechanisms is discussed. Finally, mathematical models for peripheral circuit units (crusher, cooler and screens) are presented. By coupling all the involved units, a complete granulation circuit simulator is reported. Steady-state and dynamics results obtained by means of the urea granulation simulator are provided to show the influence of different circuit operating variables on the marketable product size distribution and the plant throughput. Summarizing, this chapter gives an introduction to the main features of the urea granulation process and remarks operation problems that face the granular urea production together with possible strategies to overcome them.

NOMENCLATURE

D_{bot}	Granulator bottom diameter (m)
d_m	Arithmetic mass mean diameter (m)
$\overline{dp}_i$	Arithmetic mass mean diameter of size class *i* (m)
D_{top}	Granulator top diameter (m)
FO	Value of the objective function (-)
GAP_L	Space between lower crusher rolls (mm)
GAP_U	Space between upper crusher rolls (mm)
h_B	Bottom screen aperture (mm)

H_{CO}	Granulator conical section height (m)
H_{Cy}	Granulator cylindrical section height (m)
H_k	Height of chamber *k* (m)
h_T	Top screen aperture (mm)
H_{weir}	Weir height (m)
i	Number of a size interval (-)
k	Chamber number (-)
$M_{u\,0}$	Initial urea mass in the bed (kg)
$M_{u\,f}$	Final urea mass in the bed (kg)
$\dot{M}_{melt}$	Urea melt mass flowrate to each granulator chamber (kg/s)
O	Screen oversize mass flowrate (kg/s)
P	Screen product mass flowrate (kg/s)
R	Recycle fraction (-)
SGN	Size Guide Number (mm x 100)
$SGN_{product}$	Size Guide Number corresponding to the product stream (mm x 100)
t	Time (s)
t_0	Initial simulation time (h)
Ta_2	Second chamber fluidization air temperature (K)
t_f	Final simulation time (h)
T_k	Temperature in chamber *k* (K)
U	Screen undersize mass flowrate (kg/s)
UI	Uniformity Index
$u(t)$	Control variables (-)
v_f	Superficial velocity (m/s)
v_{mf}	Seeds minimum fluidization velocity (m/s)
w_i	Mass fraction collected within size class *i* (wt%)
$x(t)$	State variables of the process model (-)
Xu	Urea concentration of the sprayed solution ($kg_{urea}/kg_{solution}$)

Greek Symbols

α	Fraction of discharge area (-)
γ	Granulator angle (°)
η	Granulation efficiency (%)

1. INTRODUCTION

Fertilizers play a vital role to guarantee enough crops to meet the increasing population food needs and the progressive energy requirements (FAO, 2009). Among the nutrients required by plants, nitrogen, phosphorus and potassium are essential and, therefore, used in large quantities. Urea is the most important nitrogenous source consumed by the agriculture sector both as a fertilizer and animal feed additive, which makes its production considerably higher than others (Fertilizer Manual, 1998).

The twenty-first century presents many challenges for global agriculture: to produce more food and fiber to feed a growing population with a smaller workforce, to provide raw materials for a potentially huge bioenergy market, to contribute to the development of many emerging countries dependent on agriculture, to develop more efficient production methods and sustainable adaptation to climate change (FAO, 2009). Clearly, the responsible use of fertilizers is one of the keys to increase grain production worldwide. In that sense and according to the last IFA (International Fertilizer Industry Association) report, 58 new urea plants are planned to come on stream between 2010 and 2015, which would increase the installed capacity up to 224.5 Mt in 2015 (Heffer and Prud'homme, 2011). Based on historical plants operating rates and the projects with high probability of realization, the world urea supply is estimated at 190.5 Mt in 2015. Besides, the demand for urea is expected to grow from 148 Mt in 2010 to 171.7 Mt in 2015, which represents a 3.2% annual growth. About 90% of the urea demand is for use as fertilizer (Heffer and Prud'homme, 2011). In this context, knowledge improvements to operate more efficiently urea plants will be extremely worthy.

Urea, also known as carbamide (NH_2CONH_2), is a white crystalline organic chemical and the world's most common solid nitrogen fertilizer (Fertilizer Manual, 1998). It has the highest nitrogen content (46 wt%) within all the other solid sources of nitrogen available in the market. Synthetic urea is made from ammonia and carbon dioxide, which are fed to a reactor operating at high pressure and temperature. The urea is formed by two reactions in series:

$$2NH_3 + CO_2 \leftrightarrow NH_2COONH_4 \tag{1}$$

$$NH_2COONH_4 \leftrightarrow NH_2CONH_2 + H_2O \tag{2}$$

There is a third undesired reaction that lowers the urea yield and produces biuret:

$$2NH_2CONH_2 \leftrightarrow NH_2CONHCONH_2 + NH_3 \quad (3)$$

If this compound is present in urea in high concentrations (higher than 2 wt%) the crop leaves may be burned. In this situation, the product becomes unsuitable as a fertilizer (Mikkelsen, 2007). Consequently, the operating conditions of the urea reactor have to be selected to maximize the urea production. The liquid urea solution that comes out from the reaction unit is decomposed and concentrated to be further used in the production of solid forms.

Solid urea of fertilizer grade is marketed as prills or granules. Nowadays granulation is the preferred urea finishing process, basically because urea granules are far superior in quality than prills (Fertilizer Manual, 1998; Brouwer, 2010a). Prilling was the first urea finishing process developed; at present the prilling towers (see Figure 1) are still functioning and under construction. Prilled urea is made by techniques including a spinning bucket or shower heads that deliver urea droplets from the top of the tower. These droplets fall by gravity and are cooled down by an upward air flow that increases the droplets falling time and promotes the urea solidification. The urea spherical particles are collected at the bottom of the device (Mavrovic, 1976). Prilling has several disadvantages such as: fine dust formation, relatively small particle diameter (the maximum average diameter is around 2 mm) and limited prills crushing strength (Landis, 1980; Meessen and Petersen, 1996).

On the other hand, the granulation process is considered as one of the most significant advances in the fertilizers industry, overcoming the main drawbacks of the prilling process. The granulation technology purposely converts, by a sequence of events, small particles into large permanent masses in which the initial primary units are still identifiable (Kayaert and Antonus, 1997).

Typically, three components are needed to produce granules, initial seeds or nuclei, mixing and a binder. For large-scale granulation processes three types of granulators are often used: pan, drum and fluidized-bed granulators (Litster et al., 2004). In continuous urea granulators, small particles (usually called seeds) are constantly incorporated to the unit while a concentrated urea solution (frequently known as urea melt) is constantly sprayed. The solids are agitated (pans and drums: mechanical mixing; fluidized beds: pneumatic agitation) to achieve a good distribution of the binder onto the seeds surface (Litster et al., 2004). Particularly for urea granulation, the seeds enlargement occurs mainly by the solidification of droplets that successfully collide against the solids surface (Bertin et al., 2010a). The desired urea growth mechanism is often called coating. However, and as it is shown in Figure 2, depending on the relative size of droplets and seeds the growth can be achieved either by accretion or layering (Kayaert and Antonus, 1997). Accretion is preferred over layering because it gives granules with higher crushing strength.

Urea granulation is a complex operation that cannot be carried out in a single device; it is rather achieved by a combination of process units with specific functions constituting the called granulation circuit (Cotabarren et al., 2008; Cotabarren et al., 2009). The granules that leave the granulator are classified by size. Downstream, the product that has marketable granulometry is transported to storage facilities while the oversize particles are crushed and then combined with the undersize granules to be recycled to the granulator as seeds.

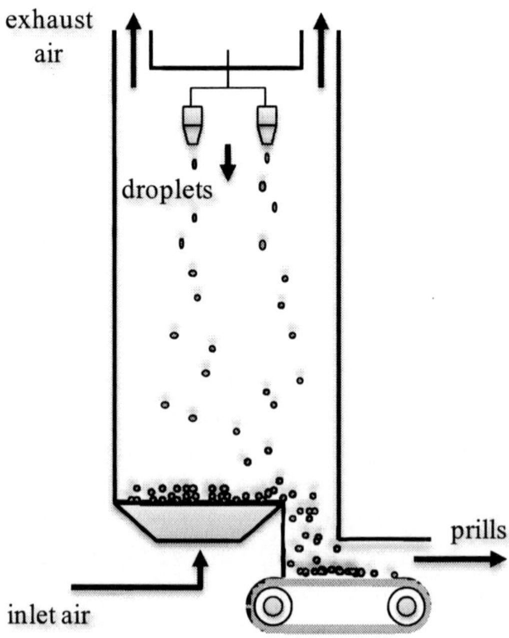

Figure 1. Prilling tower.

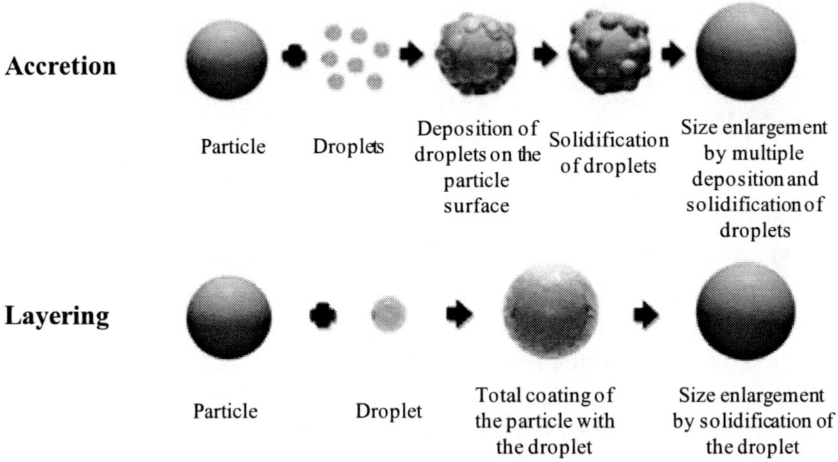

Figure 2. Desired urea size enlargement mechanisms.

The understanding of the urea granulation process is of great importance to generate granules with the desired attributes through stable operations. Although granulation is one of the fundamental unit operations in particle technology, there is still need to provide insight into the complex behavior of this process coupled with the size classification and reduction steps that occur in the integrated granulation circuit.

In this context, this chapter initiates with a brief description of the desired final product quality properties. Then, common urea granulation technologies and the corresponding flow diagrams are described. With a focus on fluidized bed granulators, the fundamental phenomena that occur in these large-scale granulation units are discussed. Although accretion is the preferred urea growth mechanism, unexpected operating situations may favor undesired size enlargement mechanisms. For this reason, and based on experimental data obtained in a pilot-scale fluidized-bed batch granulator, the influence of the operating variables on both granules quality and growth mechanisms is presented. Simple guidelines to ensure growth by coating are discussed. Finally, the peripheral circuit units (crusher, cooler and screens) are briefly described. Results obtained by using a previously developed complete granulation circuit simulator, corresponding to steady-state and dynamic sensitivity analysis, are included to show the effect of different circuit operating variables on the marketable product size distribution and its throughput. Based on the most relevant variables, as pointed out by the steady-state and dynamic simulations, an optimal control study to increase plant capacity is discussed.

2. Urea Particles Properties

The selection of the urea finishing technology should be based on, among economical and environmental reasons, the degree of accomplishment of the desired urea particles attributes (e.g., size, shape, moisture and biuret content, caking tendency, crushing strength, etc.).

The granulometry of solid fertilizers is important because particle size is related to the fertilizer dilution rate in the soil, which should be neither too high nor too low to ensure a good utilization of nutrients by crops. On the other hand, the more homogeneous in size the solid fertilizer is (i.e., the narrower the Particle Size Distribution or PSD is), the better the uniformity of spreading on the field. Low spread PSDs are quite attractive to formulate fertilizers mixtures minimizing the segregation phenomenon. Usually, the PSD of commercial urea granules is characterized by two parameters. The first one known as Size Guide Number (SGN) represents the median in millimeters of the mass distribution, multiplied by 100. In other words, the SGN is the particle size in millimeters for which 50 wt% of the sample is greater and the remaining 50 wt% is lower, multiplied by 100. The second parameter is the Uniformity Index (UI) and characterizes the product PSD spread. It is defined as the ratio between the size of an opening that allows the passage of 5 wt% of the population and the aperture size that allows the passage of 90 wt% of the population, multiplied by 100. High UI values indicate uniform PSDs. According to the international standards, the SGN and UI for granular urea are around 300 and 55, respectively (Giovanelli and Schech, 2004). To limit the product PSD spread, 90 to 95 wt% of the granules are also required to be within the size range 2 to 4 mm (www.kafcobd.com; www.cfindustries.com; Giovanelli and Schech, 2004). Granulation processes produce particles with diameters ranging from 1.5 mm to 15 mm, easily reaching the international standard (Nicks et al., 1980). On the other side, prilling leads to narrower PSDs (90 to 95 wt% of the prills between 1.2 and 2.2 mm) but with a maximum particle diameter of about 2.1 mm. Larger diameters are not viable since taller prilling towers are required, making the process more expensive and operationally unstable (Brouwer, 2010a).

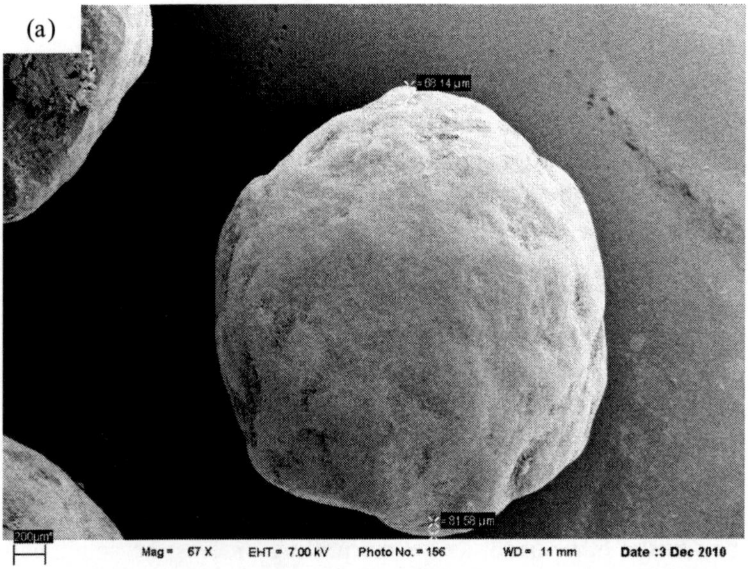

Figure 3.a. SEM micrographs of commercial urea granules: superficial morphology.

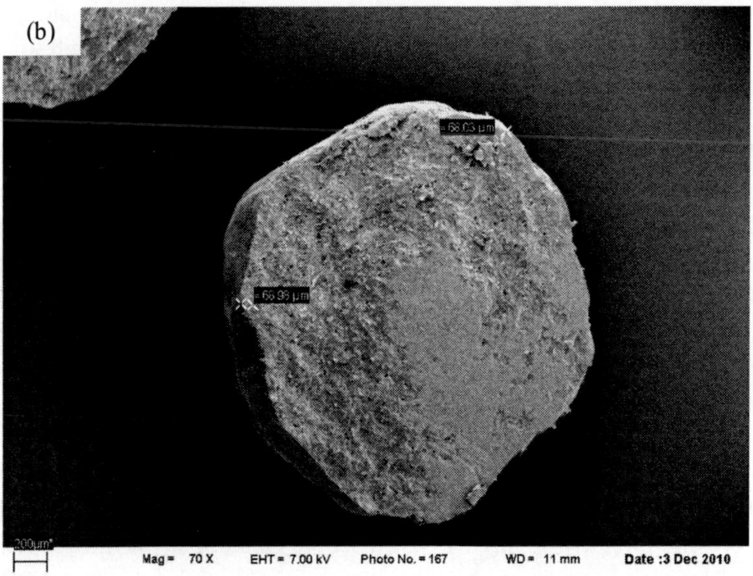

Figure 3.b. SEM micrographs of commercial urea granules: internal morphology.

A fertilizer with good flowability will flow smoothly (Prescott and Barnum, 2000). In contrast, poor flowability tends to consolidate granules during storage and makes difficult the material handling. Flowability is favored by large, smooth and spherical particles with low humidity and uniform particle size (Meessen and Petersen, 1996). Figure 3.a shows the SEM micrograph of a commercial urea granule, the particle is quite rounded which is, as mentioned, one of the pursued product attributes to improve flowability.

The urea particles density depends on the degree of consolidation reached during their growth. When the main growth mechanism is coating, the solids mixing within the granulators favors the particles consolidation and, consequently, the production of very low porous structures. For example, by using fluidized-bed units, the urea granules have densities very close to the solid density of 1333 kg/m^3 (Daubert and Danner, 1996). Due to the low internal porosity, urea granules are more resistant to breakage and attrition than prills. Figure 3.b confirms that the commercial urea granules have low porosity. On the other hand, as for the prilling process the solidification by cooling of the urea melt droplets takes place from the outside to the center, internal cavities are created giving particles with higher porosities and, thus, lower crushing strength than granules. For urea bulk handling and storage, relatively high crushing strengths are required. The crushing strength for 3 mm urea granules is about 4 kg_f while for 1.7 mm urea prills is around 1.2 kg_f (Brouwer, 2010a).

As aforementioned, biuret is an undesired compound in urea fertilizers (Mikkelsen, 2007). During urea finishing processes, urea may decompose into different byproducts including biuret. The decomposition is favored when the urea is treated even slightly above its melting point and/or for long residence times (Finck, 1988; Kayaert and Antonus, 1997). Both prilling and granulation usually give solid urea particles with biuret contents below the maximum admissible one, around 0.70 - 0.85 wt% (Snamprogetti-Saipem, 2007).

Hygroscopicity is the degree to which a material will absorb moisture from the atmosphere. All fertilizers are soluble in water to some extent and their hygroscopicity is important when considering conditions for bulk storage, handling and field application. The fertilizer caking problems are expected to be more important for highly hygroscopic materials. Hygroscopicity is affected by many factors such as moisture content, ambient temperature and relative humidity, particle structure and porosity, pressure and storage time. It has been shown that the tendency to caking is decreased if the fertilizer moisture content is kept as low as possible, the product is stored at low temperature and under the lowest feasible outside atmosphere humidity (Kars,

1984). The urea lumps formation can also be minimized if the granules are relatively big and, for bulk storage, if the pile-static forces (pile height) are reduced (Sharma and Patel, 2000). Therefore, granulation processes are designed to obtain almost complete dried and cool enough particles to avoid lumps formation. The moisture content of commercial urea granules is about 0.2 wt% while for urea prills it is between 0.15 and 0.30 wt% (www.uhde-fertilizer-technology.com; Brouwer, 2010a).

Small amounts of formaldehyde are frequently added to the urea melt, before prilling or granulation takes place, with the following purposes: to diminish dust formation during the urea finishing and/or handling processes, as anti-caking agent and/or to improve the product crushing strength. This particular additive decelerates urea crystallization within the granulation unit, allowing the outer layer of the particle to contain a relatively high fraction of liquid urea (free of water) for some time. During that time, the outer layer remains plastic increasing the resistance to dust formation and, thus, improving the final product quality (Kayaert and Antonus, 1997; Niks et al., 1980). The formaldehyde content in commercial urea granules is typically between 0.30 and 0.55 wt%, while for prills it is in the range 0.10 to 0.30 wt% (Brouwer, 2010a).

The prilling process has been extensively used in the past because of its simplicity and low cost. Many urea plants in the world operate with prilling towers, being those several decades old (Brouwer, 2011). However, as it can be inferred from the properties of the solid forms above discussed, nowadays the granulation process allows the production of granules with higher quality, being the average properties summarized in Table 1.

Table 1. Urea granules properties

Property	Urea Granule
SGN	~ 300
UI	>55
Mass fraction between 2-4 mm, wt%	90-95
Density, kg/m^3	1333
Crushing strength (3 mm), kg_f	4
Biuret content, wt%	0.7-0.85
Moisture content, wt%	0.15-0.30
Formaldehyde content, wt%	0.3-0.55

3. Granulation Technologies

Many granulation techniques have been developed and applied to overcome prilling drawbacks, including pan granulation (e.g., by Tennessee Valley Authority-TVA and Norsk Hydro), drum systems (e.g., by C&I Girdler, Kaltenbach-Thuring and Montedison), fluidized-bed processes (e.g., by NSM, next Hydro-Agri, later Yara, today Uhde Fertilizer Technology-UFT and Stamicarbon) and spouted-bed technologies (by Toyo Engineering Corporation-TEC) (Baeder, 2010; Fertilizer Manual, 1998).

In the early days, and despite of the capacity limitation (typically up to 500 tons per day, tpd), the drum granulation processes introduced in the 1960's were commonly used (Brouwer, 2010a). Later, the fluidized-bed and spouted-bed granulations developed during the period 1970-1980 became more popular techniques (Buitnik, 2009). Today, fluidized-bed granulation is the leading urea finishing technology because it offers single-train units which are according to the current urea plant sizes (Fertilizer Manual, 1998; Uhde Technology Profile, 2002).

Economy of scale has been driving urea plant capacities higher and higher, from 70 tpd in the 1960´s to 3000 tpd by the 2000´s and nearly 4000 tpd currently (Brouwer, 2010b). Nowadays, there are seven large-scale single line urea plants (defined as urea plants which are able to produce a minimum of 2800 tpd) in operation (Brouwer, 2010b). These plants are summarized in the first seven rows of Table 2.

As it can be seen, six of them were licensed by Stamicarbon and the remaining one by Saipem. Regarding the urea finishing process, just one is based on prilling while the others are urea granulation plants. Except for Profertil, which is constituted by two UFT fluidized-bed units in parallel (design capacity of 1850 tpd each), all granulation plants in operation are single line UFT fluidized-beds.

In addition, seven Stamicarbon and five Saipem large-scale single line urea plants have been recently started or are expected to start production in the near future. As it is reported in Table 2, eight of these twelve plants were planned to include a single line UFT fluidized-bed granulation section, three of them a Stamicarbon fluidized-bed granulator while just only one was designed based on prilling as finishing process.

Besides, in the latest Worldwide Urea Plants Overview, Wang and Brouwer (2011) reported a total of about 460 urea plants in operation with an overall actual plant capacity of around 175 million tpy, tons per year, (i.e., 480000 tdp).

Table 2. Actual and next large-scale single urea plants (Brouwer, 2010b)

Licensor	Licensed/ Completion	Company	Plant Site	Finishing Process*/ Technology§	Capacity, tpd
Stamicarbon	1996/1996	Yara Canada‡	Canada	UG/ SL UFT FB	2850
Saipem	1998/2001	Profertil‡	Argentina	UG/ Two UFT FB plants	3250
Stamicarbon	2001/2004	Qafco 4‡	Qatar	UG/ SL UFT FB	3500
Stamicarbon	2001/2008	Pardis 1‡	Iran	UG/ SL UFT FB	3500
Stamicarbon	2003/2007	Safco 4‡	Saudi Arabia	UG/ SL UFT FB	3600
Stamicarbon	2004/2010	Pardis 2‡	Iran	UG/ SL UFT FB	3250
Stamicarbon	2005/2005	Erdos‡	China	UP/TEC	3520
Stamicarbon	2007/2010	Unichem Sorfet	Argelia	UG/ SL UFT FB	3450
Stamicarbon	2008/2011	Lordegan	Iran	UG/ Stamicarbon FB	3250
Stamicarbon	2008/2011	Golestan	Iran	UG/ Stamicarbon FB	3250
Stamicarbon	2008/2011	Zanjan	Iran	UG/ Stamicarbon FB	3250
Stamicarbon	2008/2011	Yara Canada	The Netherlands	UG/ SL UFT FB	3500
Saipem	2008/Under construction	Hengam Petro-chemical Company	Iran	UG/ SL UFT FB	3250
Saipem	2008/Under construction	Engro Chemical Pakistan Limited	Pakistan	UP	3835
Saipem	2008/2011	Qafco 5	Qatar	UG/ SL UFT FB	3850
Stamicarbon	2009/2013	Ruwais	Abu Dhabi	UG/ SL UFT FB	3500
Saipem	2009/2012	Qafco 6	Qatar	UG/ SL UFT FB	3850
Saipem	2009/2012	Algeria Oman Fertilizer	Algeria	UG/ SL UFT FB	2x3500
Stamicarbon	2010/2014	Petrobras	Brasil	UG/ SL UFT FB	3600

(*) UG: Urea Granulation; UP: Urea Prilling. (§) SL: Single Line; FB: Fluidized Bed.

(‡) Plants currently in operation.

On this basis, the above-listed nineteen large-scale urea plants represent nearly 5% of all urea plants around the world and almost 15% of all the actual plant capacity (i.e., around 25 million tpy, 70000 tpd). It is interesting to remark that seventeen of these nineteen large-scale urea plants operate with fluidized-bed granulation technologies as finishing process. Moreover, except for three, all these plants include the particular UFT fluidized-bed granulation section.

Furthermore, and with respect to the overall production capacity in operation, UFT shares today more than 80% of the market for fluidized-bed urea granulation technologies with a maximum single-train capacity of 3850 tpd (www.uhde-fertilizer-technology.com).

The UFT fluidized-bed granulation technology was developed in the mid 1970's by NSM (then Hydro-Agri, later Yara, today Uhde). The process was first successfully implemented on an industrial plant with 800 tpd in Sluiskil (The Netherlands) in 1979 (Niehues and Antonus, 2006). Almost simultaneously (from 1977 to 1983), the Stamicarbon fluidized-bed granulation was also developed. However, this technology was not commercialized until recently. Indeed, it was first implemented in a commercial pilot scale (280 tpd) in 2002 in cooperation with Grodno Azot (Meessen and van Baal, 2003) and in an industrial scale during 2003 for Agrium. Since then, Stamicarbon's fluidized-bed granulation technology has been licensed over 15 times and with capacities up to 3900 tpd for plants on stream or under execution (Buitnik, 2009; www.stamicarbon.com). As it is discussed in the following section, the main difference between these two fluidized-bed granulation processes is related to the granulator spraying technique.

On the other hand, in the 1970's a spouted-bed granulation technology was introduced by Toyo Engineering Corporation (TEC). In late 1980's, TEC further improved its granulator design by applying a spouted-fluidized bed technology, which is a combination of spouted and fluidized beds. Since 1975, TEC has licensed 18 urea spouted-fluidized bed granulation plants, scaling up from 50 tpd in 1975 to 3250 tpd in 2010 (Nakamura, 2007).

The granulation unit and process flows of the three abovementioned urea granulation technologies are presented below in a comparative manner.

4. Granulation Unit

The urea world capacity information, briefly presented in the previous section, clearly indicates that the UFT fluidized-bed granulation technology is today the most popular one for granular urea production. Therefore, in this section a detailed description of the UFT fluidized-bed granulator is given, stressing the main differences with respect to the two other aforementioned common fluidized-bed granulation units (i.e., Stamicarbon fluidized-bed and TEC spouted-fluidized bed granulators).

Figure 4 shows a simplified scheme of the UFT fluidized-bed granulator, which is constituted by several fluidized bed chambers in series. The first chambers are for particle growth while the last ones for granules conditioning, polishing and cooling (Kayaert, 1980). Each bed of granules is supported on a perforated plate through which the fluidization air flows up. In the growth chambers, a 95-97 wt% urea solution at 130-132 °C is atomized upwards onto the fluidized particles by air-assisted two-fluid nozzles located just above the perforated plate (Franzrahe, 2010; Potthoff, 2007; www.uhde-fertilizer-technology.com).

The atomizing air is supplied at a 1.0-1.2 barg and 130-135 °C, spraying droplets around 20 μm (Alnajar, 2010; Kayaert, 1980). Formaldehyde, as mentioned before, is added to the urea solution (around 0.45-0.6 wt%) as granulation additive to reduce dust formation and as anti-caking agent to improve storage properties (Alnajar, 2010; Kars, 1984). As the particles move through the growth chambers that operate at 109-112 °C, they grow by accretion with simultaneous urea solidification and water evaporation (Alnajar, 2010; Kars, 1984; Franzrahe, 2010).

The licensing company Uhde Fertilizer Technology claims that the UFT product is a hard granule with low moisture content, far superior in quality than granules produced through layering processes. The urea solidification is accompanied by the release of a large amount of heat. Due to the relatively low urea melt concentration (i.e., 95-97 wt%) with respect to that of other technologies, a higher quantity of the heat produced as the urea solidifies is removed by the evaporating solution water. This reduces the amount of fluidization air needed in the granulator (approximately 88 Nm^3/h of fluidization air per tpd of granular urea; Alnajar, 2010) to keep the growth chambers temperature at the required level (109-112 °C). Suspended solids move from one chamber to the following one, underflowing through slots between the perforated plate and the partitioning walls (also known as weirs). In the cooling chambers (without spray nozzles), the particles are cooled down to 70-90 °C (Niks et al., 1980). The granules leave the granulator by bottom flow while the seeds (recycle of granules out of specification) are fed in the first chamber above the bed surface, which height is around 0.7-0.75 m (Alnajar, 2010; Kayaert, 1980). The air space over the bed is kept under slight vacuum (-0.041 to -0.046 barg) to avoid dust in the surrounding area (Alnajar, 2010).

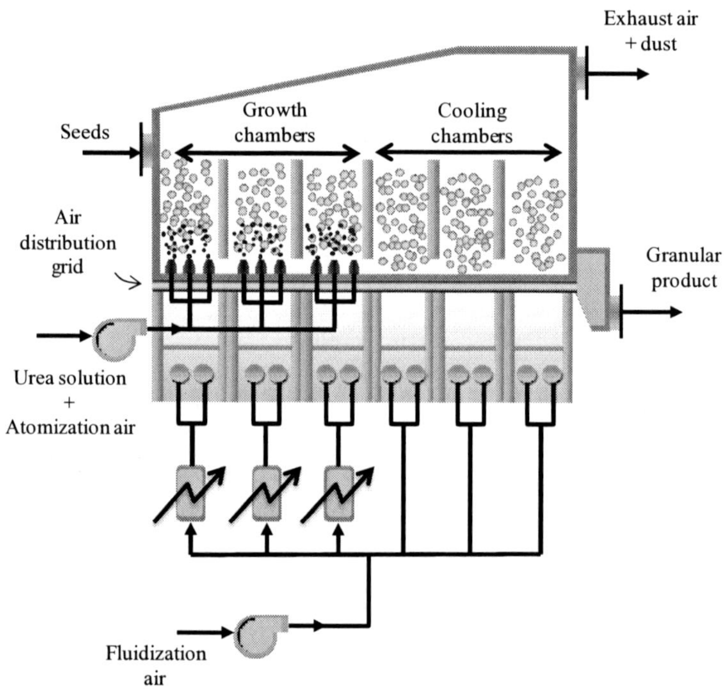

Figure 4. Schematic of a UFT fluidized-bed granulator.

The Stamicarbon fluidized-bed granulator is essentially the same device, being its key feature the low pressure film spraying nozzles used to supply a highly concentrated urea melt (i.e, 98.5 wt% urea solution) at around 140 °C (Stamicarbon, 2011; Meessen and van Baal, 2003; www.stamicarbon.com). As it can be seen in Figure 5, each injection header comprises vertically placed risers fitted with spray nozzles that spray the urea melt as a film instead of droplets. The headers are also connected to a secondary air stream, which is supplied through an annulus around the melt sprayer at high velocity creating a zone with reduced pressure at the top of the nozzle. As a result of the lower pressure, the particles are sucked into the urea melt film (Roos, 2008). Each time the particles pass through the liquid film they are repeatedly covered with a thin layer of urea melt, being the growth mechanism layering instead of accretion (Stamicarbon, 2011; www.stamicarbon.com).

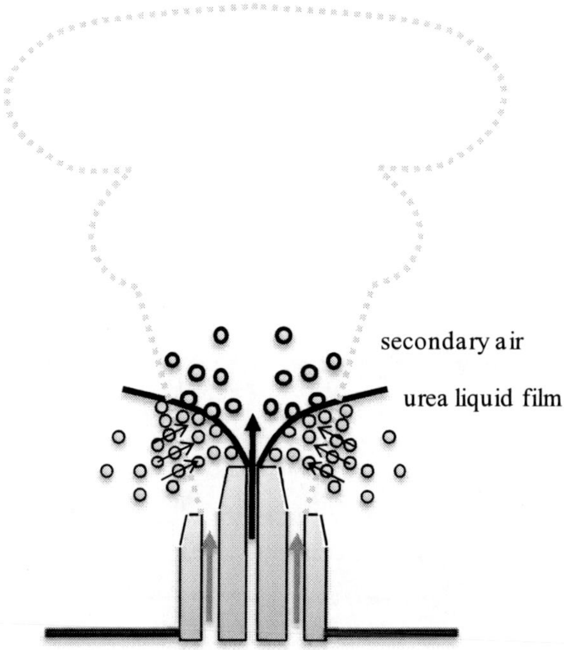

Figure 5. Schematic of a low pressure Stamicarbon film spraying nozzle.

Consequently and according to the licensing company, the particle size growth is homogeneous and progressive, resulting in uniform shape and good quality granules. These particular sprayers reduce the amount of dust formed as compared to high pressure spraying or two-fluid nozzles (as the ones used in the UFT technology), in which the urea is atomized as a fine mist (Brouwer, 2010b; Meessen and van Baal, 2003). In addition, a minimum amount of formaldehyde (about 0.3 wt%; Buitnik, 2009) is required. In fact, this compound is injected to the urea melt just to stabilize the final product (Roos, 2008). The Stamicarbon granulation chambers typically operate at the same thermal level than those of the UFT design (approximately 110 ºC), being the temperature at the outlet of the granulator cooling section roughly 95 ºC (www.stamicarbon.com).

Both, the UFT and Stamicarbon fluidized-bed granulators have been scaled-up through the years by optimizing compartment size or adding additional compartments to obtain the required granulation throughput.

On the other hand, the TEC industrial multi-stage spouted-fluidized bed granulator consists of several spouted beds in a fluidized bed that is formed above the perforated plate surrounding the spouted beds (see Figure 6). Each spouted bed has only one pressure spray nozzle that sprays a urea solution of 94-98.5 wt% urea concentration into droplets (Sakata et al., 2007; www.toyo-eng.co.jp). As no compressed atomizing air is required, the power consumption is reduced.

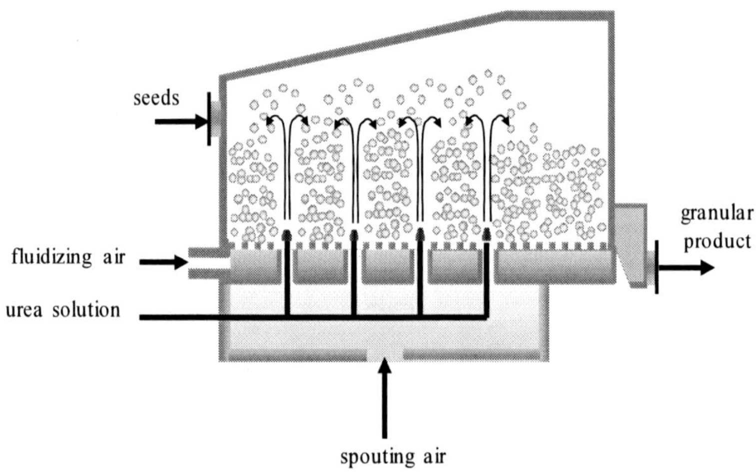

Figure 6. Schematic of a TEC spouted-fluid bed granulator.

Table 3. Urea granules properties as reported by the licensing companies

Property	UFT fluidized-bed granulation	Stamicarbon fluidized-bed granulation	TEC spouted-fluidized bed granulation
Moisture content, wt%	0.2	0.25	0.25
Biuret content, wt%	0.7-0.8	0.8	0.7
Formaldehyde content, wt%	0.4	0.2	0.45
Crushing strength, kg_f	4.1 (3 mm)	3 (2.8 mm)	3.5 (3 mm)
Size distribution wt% (2-4 mm)	95	95	95

The particles, suspended by the spouting air, grow gradually while passing through the spouted beds. The vigorous mixing in the spouted bed gives round and uniform granules. Besides, optimal spouted bed air velocity minimizes dust formation in the granulator. This granulator also operates under minor negative pressure but at a slightly higher thermal level than the UFT and Stamicarbon fluidized-bed units. The TEC operating temperature is controlled in a range of 110-120 ºC to accelerate drying of the relatively low urea concentration solution without affecting urea granulation and giving granules with 0.3 wt% moisture content or less (Nakamura, 2007). According to Sakata et al. (2007), a higher fluidized bed temperature has a higher drying effect. However above 120 ºC the urea solution droplets deposited onto the particles surface do not instantly solidify. Slow urea solidification gives a granular product with lower crushing strength (Sakata et al., 2007). On the other hand, under 110 ºC the drying is not notoriously improved by increasing the fluidized bed temperature. The temperature of the particles leaving the granulation unit is similar to that of the two above described fluidized-bed technologies. Indeed an after-cooling zone, located near the granulator exit, cools down the granules to about 90 ºC. The TEC granulator has been simply scaled-up increasing the number of spouted beds (www.toyo-eng.co.jp). To summarize, Table 3 presents typical quality of the urea granules produced by the UFT fluidized-bed granulation, Stamicarbon fluidized-bed granulation and TEC spouted-fluidized bed granulation.

5. Granulation Circuit

Contrary to prilling (where solid material of the required final size is obtained by direct solidification of relatively big molten urea droplets), all granulation processes are based on solidifying successive tiny droplets or layers of urea solution or melt onto the surface of particles (known as seeds) that are separately fed into the granulator and grow during their passage through the granulator by continuous spraying (Kars, 1984). Since growth is not uniform, the stream of solids from the granulator is generally classified by screening into three fractions: oversize material (which is frequently crushed and then recycled back to the granulator), final product and undersize material (which is directly recycled to the granulator). This requires a recycle system, which is usually not necessary for prilling. The whole system, frequently called granulation circuit, mainly consists of coolers, screens, bucket elevators and crushers.

Despite the differences in granulator design, the UFT, Stamicarbon and TEC granulation circuits have similar flow schemes as far as other granulation processes available in the world (www.toyo-eng.co.jp). Actually, for the same unit capacity, product grade, recycle ratio (expressed as recycled material to on-specification granules) and quality of installed equipment, it can be assumed that the operation and on-stream time of this plant section are almost independent of the involved granulation technology (Kars, 1984).

In this section, and for the reasons stated above, the UFT fluidized-bed urea granulation circuit is described in detailed pointing out the differences with the circuits associated to the Stamicarbon and TEC granulation technologies.

Figure 7 shows the flow diagram of the UFT fluidized-bed urea granulation circuit. The heart of the plant is the UFT fluidized-bed granulator. The product discharged from this central unit flows by gravity to the first fluidized-bed cooler and thereafter is taken up to the screening section by a bucket elevator. In the double-deck vibrating screens the urea granules are separated into the above mentioned three fractions. The undersize is directly recycled into the granulator while the oversize is recycled after crushing in double-roll crushers. According to Alnajar (2010), the gap of the upper pair of rolls is ideally about 50% of the top deck screen (oversize mesh size) whereas the gap of the lower pair of rolls is around 50% of the bottom deck screen (lower size mesh). The mesh sizes are determined to meet the final product requirements. The on-size stream passes through the final fluidized-bed cooler, where it is cooled down to around 40 °C, and is then transported to product storage (Alnajar, 2010). For the final cooler there is another option, which consists in a bulk flow cooler that is operated with cooling water instead of air (Potthoff, 2007). Cooling the urea to sufficiently low temperatures is a key issue to avoid caking in the bulk storage warehouse. In UFT's process, the solids recycle ratio ranges from 0.5 to 1 (UFT Brochure, 2011).

The Stamicarbon fluidized-bed granulation circuit is analogous to that of the UFT fluidized-bed technology, being the granules temperature at the outlet of the first fluidized bed cooler around 70 °C (Roos 2008; Stamicarbon, 2011).

On the other hand, TEC urea granulation process is similar but no identical. In contrast to the UFT and Stamicarbon fluidized-bed granulation circuits, the TEC flow diagram does not include a cooling unit downstream the granulator. As a result, the crushed material and undersize particles from the screen are recycled back to the granulator at moderately high temperature (about 90 °C). This seed circulation at moderately high temperature minimizes the cooling air requirement downstream the granulator, which operates at relatively high temperatures (i.e., 110-120 ºC) in comparison to other technologies. Like in the UFT and Stamicarbon granulation, the granules on specification are further cooled in a product cooler to temperatures below 60 ºC (fluidized-bed cooler or water bulk flow cooler) before being sent to storage (Nakamura, 2007).

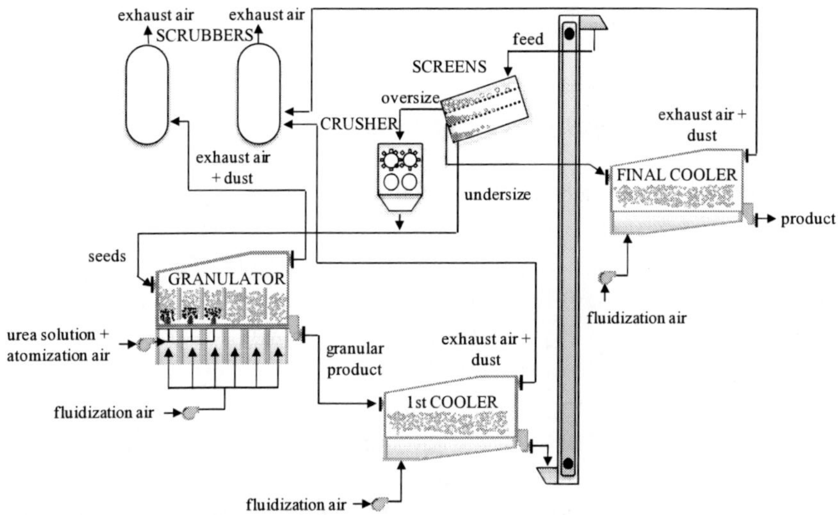

Figure 7. Process flow diagram for the UFT fluidized-bed urea granulation circuit.

To match current urea plant sizes, the solids screening, crushing and handling operations have been usually scaled-up through the years by simply adding the required number of bucket elevators, screens and crushers considering the capacity of the available units.

6. CHALLENGES IN FLUIDIZED-BED GRANULATORS MODELING

As mentioned in the previous section, the fluidized-bed granulator is the central unit of the granulation circuit. Despite the widespread use and large-scale application of the granulation process in the fertilizers industry and the significant advances in understanding the fundamentals of granulation during the last two decades, the granulation units are still operated mainly by trial and error. Therefore, it is important to understand more deeply the granulator behavior in order to develop tools for maximizing the plant profit while simultaneously obtaining a marketable product. Accurate mathematical models should help to operate and scale-up granulation units rationally. In that sense, in this section the size change mechanisms that can occur during the granulation, the binder solidification phenomenon, different approaches to model fluidized-bed granulators and some selected simulation results of a fluidized-bed urea granulator are presented.

6.1. Particle Size Change Mechanisms

The granulation processes are usually classified according to the binder nature as wet, dry o melt. Wet granulation refers to the use of an enlargement agent that is dissolved or suspended in an easily evaporable solvent (usually water). In dry granulation, fine solid particles are added to the seeds bed, the adherence of the powder is promoted by van der Waals or electrostatic forces. Melt granulation can be carried out via two different procedures. The first method consists in spraying a hot melt agent within a bed of cooled particles, while the second one is based on the addition of a powder binder that is heated in the granulator up to a temperature close to its melting point. Thus, the powder binder gets soft and spreads over the seeds. The binder deposited layer solidifies by further cooling (Saleh and Guigon, 2007). The urea granulation involves the use of highly concentrated aqueous solutions, therefore to some extent the binder nature is much closer to molten agents than those used in wet granulation. In what follows, the urea process will be described as liquid granulation.

Throughout the whole granulation process, the particles are subjected to many size enlargement or reduction mechanisms (see Figure 8) that define the final particle size distribution of the population. Depending on the specific process, those mechanisms can occur alone or simultaneously (Bucalá and Piña, 2007; Litster et al., 2004; Tan et al., 2006).

As previously described, granulators are based in size enlargement processes to achieve the desired particle growth (Cameron et al., 2005). The granules size enlargement involves the addition of the binder, which for the urea process is fed in the form of liquid droplets (or film in the special Stamicarbon fluidized-bed granulation technology). The first stage in the growth process is the deposition of the droplets onto the particles surface. When the spray solution droplets successfully reach the particles, two types of size enlargement processes can occur: coating (accretion or layering) or agglomeration (Smith and Nienow, 1983).

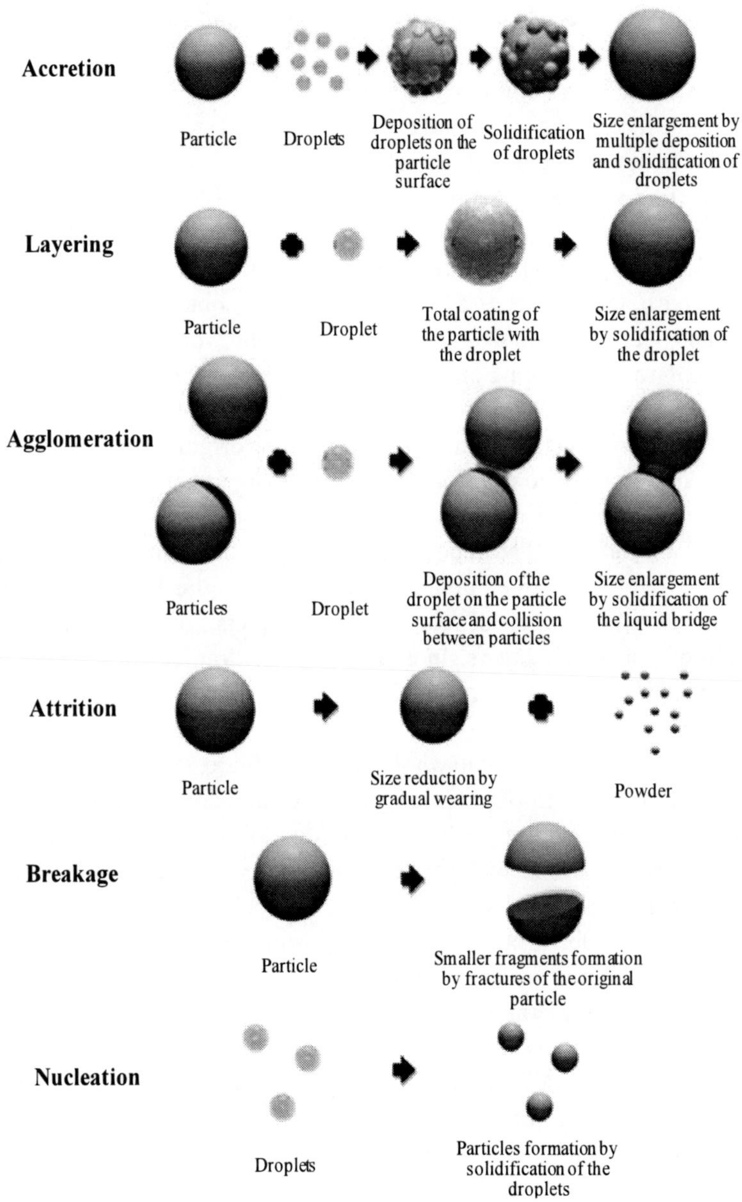

Figure 8. Mechanisms of particles size change.

Coating does not change the particles number of the population; instead an increase in the population mass or volume occurs (Bucalá and Piña, 2007). Agglomeration is a size enlargement mechanism that competes with coating. In agglomeration, two or more particles collide and stick (see Figure 8) (Pietsch, 1991). This process is facilitated by the liquid binder. The liquid binds the particles by a combination of capillary and viscous forces (Iveson et al., 2001). Permanent bonds are formed during subsequent drying/solidification. The possibility of two particles to aggregate is dependent on many factors such as the strength and deformability of the agglomerates and the availability of liquid in the proximity of their surfaces (Ennis et al., 1991; Tardos et al., 1997). Unlike coating, agglomeration decreases the total number of particles. Figure 9 shows urea granules obtained in a pilot-scale fluidized-bed urea granulator using different operating conditions, corresponding to pure coating and agglomeration growth mechanisms (Figures 9.a and 9.b, respectively).

The relative size between the droplets and the original particles is one of the factors that determine the dominant size enlargement mechanism. If the droplet size is of the same order as the particle size, growth occurs by layering. In this case, the droplet that deposits onto the particle completely covers the seeds surface with a layer of liquid material (see Figure 8). The subsequent deposition and solidification of relatively big droplets produces granules with a structure of concentric layers. In contrast, if the droplets are much smaller than the particles, the coating mechanism is called accretion. In this case, each droplet covers a small portion of the total particle surface and a large amount of droplets is required to coat the whole particle (Figure 8).

The crushing strength of granules grown by accretion has been reported as higher than that of those grown by layering due to the important stresses that inherently develop in each layer for onion-like structures (Kayaert and Antonus, 1994).

Besides the binder droplet size, coating and agglomeration mechanisms depend on other factors such as the binder viscosity or the binder/particles mass ratio. The agglomeration occurs when liquid bridges between particles (two or more) are formed, which solidify because the process disruptive forces (e.g., agitation) are insufficient to achieve the granules separation (Turton et al., 1998). Agglomerates have a non-homogeneous structure and considerably lower quality (moisture content is usually higher, flowability is poorer, there is not an efficient control of particle size, etc.) than granules produced by coating (Capes, 1980). Even for small droplet sizes, agglomeration is a possible mechanism but can be diminished by increasing agitation in the system. In the case of granular urea, agglomeration should be avoided or minimized being accretion the desired growth mechanism. Furthermore, the occurrence of agglomeration can significantly affect the stability of the urea granulator operation leading to partial or complete defluidization and, in extreme situations, to unscheduled plant shutdowns.

Figure 9.a. Urea granules obtained by coating in a pilot-scale fluidized- bed granulator.

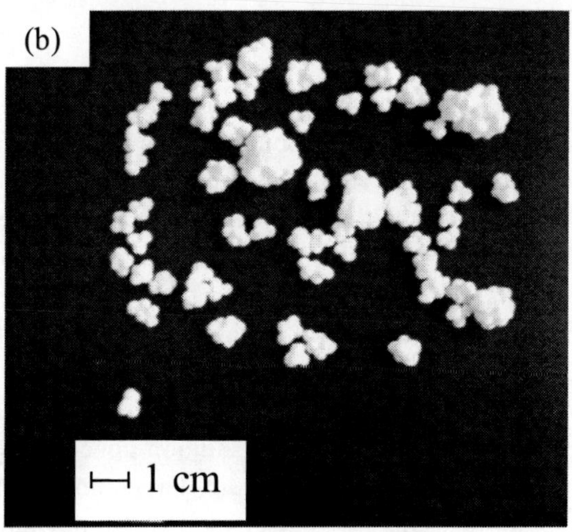

Figure 9.b. Urea granules obtained by agglomeration in a pilot-scale fluidized- bed granulator.

Attrition is the generation of dust by wearing of particles surface, which causes a gradual size reduction (see Figure 8). Although attrition can take place in the whole granulation circuit, it is more likely within fluidized-beds (i.e., granulator and cooler). In these units, urea particles are in vigorous motion and, thus, inevitably subjected to mechanical stress due to interparticle collisions and impacts with the chamber walls (Werther and Reppenhagen, 2003). Attrition generates fines so small in size that are not considered particles. Hence, attrition does not change the population number but decreases the total mass gradually as the particles shrink. Within the granulator, the interparticle collisions near the spray jet increase the attrition rate (Cruz et al., 2010). Therefore, attrition is particularly important when the spraying system is located at the bottom of the chambers (i.e., bottom spray granulation). To avoid severe attrition of the urea particles, lower fluidization air velocities are required although low levels of agitation can lead to agglomerates formation. Breakage is an improbable mechanism in the granulator, but is relevant in the granulation circuit crusher.

Finally, nucleation is the birth of particles by solidification of droplets (see Figure 8). Unlike coating, agglomeration, attrition and breakage mechanisms, nucleation is not a particle size change phenomenon but a particles generation process (Bucalá and Piña, 2007).

6.2. Binder Solidification Phenomenon

In urea UFT fluidized-bed granulation, as mentioned, the binder behaves almost like a melt due to the high concentration of the urea solution (95-97 wt%) (Kayaert and Antonus et al., 1997; Nijsten and Starmans, 1998; Niks et al., 1997). In fact, most of the urea present in the droplets solidifies by cooling. This can be explained in terms of the urea solubility as a function of temperature, which is shown in Figure 10. The selected temperature range covers the thermal changes expected for the urea solution within the growth chamber of a UFT fluidized-bed industrial granulator (Nijsten and Starmans, 1998; Niks et al., 1997). Point A in Figure 10 represents the typical urea solution inlet conditions (≈ 96 wt% urea solution concentration and 132 °C, Fertilizer Manual, 1998). At this temperature, the composition in urea is less than its water-solubility. Due to the small size of the atomized droplets (Kayaert and Antonus, 1997), a sharp decrease in the temperature of the droplets sprayed into the chamber is expected (i.e., from 132 °C down to the fluidized-bed temperature that is about 100 °C). However, when the solution reaches approximately 122 °C (point B) it becomes saturated. Consequently, the additional cooling from this temperature to 100 °C (point C) causes the solidification of about 70% of the urea present in the binder with a simultaneous increase in the water content of the solution. Besides and because of the fast decrease in the tiny droplets temperature, it can be assumed that the water evaporation initiates once the urea solution is at point C of Figure 10. At this state, the water evaporation causes precipitation of the remaining urea (about 30%).

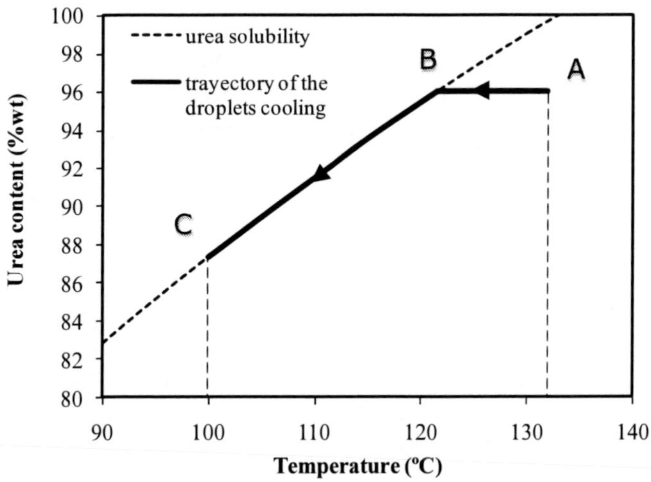

Figure 10. Urea solution droplets cooling trayectory in a fluidized-bed granulator growth chamber.

In the urea granulator, different thermal phenomena are involved. In addition to the sensible heats associated to the streams that enter and leave the granulator, latent heats are present in the heat balances. When urea solidifies, a large amount of heat (the dissolution one) is released. On the other hand, the water evaporation from the urea solution withdraws energy from the system balancing to some extent the exothermic dissolution heat. The evaporation heat per mass unit is about nine times higher than the urea dissolution one. Nevertheless, due to the high urea concentration of the solution (about 96 wt%), the total dissolution heat results much higher than the total evaporation one. Regarding the sensible heats, the seeds and fluidization air streams remove heat from the chambers while the urea melt constitutes a source of heat.

6.3. Modeling of Fluidized-Bed Granulators

6.3.1. Mechanistic Models

Granulation processes (such as the granular urea production) have been modeled with different degrees of complexity. The usual approach is based on conservation equations which are formulated to link the variables of interest. These equations are closed with constitutive relations obtained from empirical information or theoretical correlations. Thus, mass, momentum and energy balances are solved to estimate holdups, flowrates, temperatures, pressure drops, etc.

However, by means of mass, momentum and energy balances it is not possible to calculate the PSDs. The population balance equation (PBE) must be considered in particulate system modeling (Ramkrishna, 2000). The PBE is the formal tool to mathematically describe the changes in properties of the population of particles during processes such as granulation, crystallization, comminution, etc. (Randolph and Larson, 1971; Ramkrishna, 2000). For the granulator (Figure 11), the PBE solution allows computing the PSD of the solid stream that leaves the granulator (Randolph and Larson, 1971; Saleh and Guigon, 2007). Although there are analytical solutions for some special cases of the PBE, in general numerical methods are required to calculate de outlet PSDs.

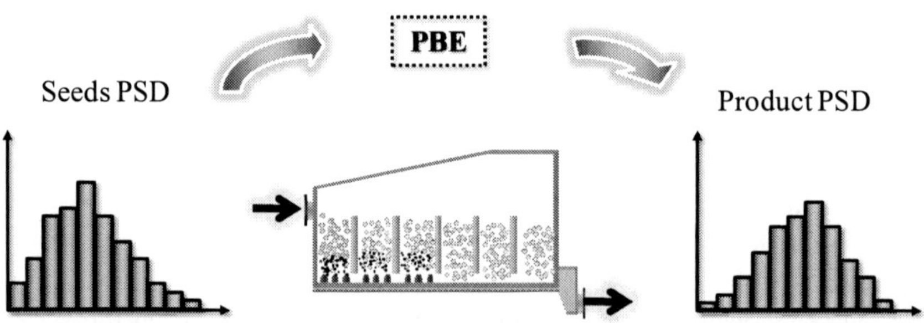

Figure 11. The role of the PBE in predicting the product PSD.

Different studies for fluidized-bed granulators with different scales and geometries have been performed; most of them were done for non-porous and spherical glass beads. Heinrich and Mörl (1999) developed a mathematical model based on mass and energy balances for the solid, liquid and gas phases to represent a fluidized-bed granulator operating both in batch or continuous mode. The binder used to produce the particle growth was a dilute aqueous solution. By means of simulations, the air humidity and air and liquid temperature profiles as a function of spatial coordinates and time were studied. The air temperature was found to be almost uniform within the bed due to the high heat transfer (except in the region near the distributor). The particles temperature was almost constant due to the high mixing degree. In later works, Heinrich et al. (2002) solved the non-steady state PBE for a continuous fluidized-bed granulator with classified discharge (i.e., the particles leave the granulator according to their size and the outlet flowrate) and external separation (i.e., simulating a granulation circuit consisting of a granulator, screens and crusher, the latter two units were represented by very simplified models). The following model assumptions were considered: constant bed mass (holdup), a binder consisting of a very dilute aqueous solution, a population of particles with size and density distributions, pure coating growth and the possibility of attrition by means of a net growth rate. Heinrich et al. (2005) coupled the PBE to the mass and energy balances for the continuous fluidized-bed granulator, assuming the air in plug flow, axial and radial mixing of solids and a homogeneous fluidization regime.

Nagaiah et al. (2008) reported a three-dimensional (time, radius and height of the bed) model to simulate the coating process in a continuous fluidized-bed granulator. They calculated the droplets concentration in the spray zone, which determines the liquid temperature and wetting efficiency. This work is similar to that published by Heinrich and Mörl (1999) regarding the constitutive equations used for modeling, but uses more sophisticated numerical methods for solving the mathematical model of granulators with different geometry. The authors successfully validated the proposed model with experimental data by spraying a dilute solution on glass beads. The results indicated good mixing of the fluidized bed. Although the influence of the nozzles on the axial profiles was important, Nagaiah et al. (2008) considered that the fluidized bed could be assumed isothermal.

Previous research suggests the existence of different thermal zones in fluidized-bed granulators. Thus, to analyze the coating mass distribution in a discontinuous granulator, Ronsse et al. (2007) horizontally divided the fluidized bed of a top-spray granulator into several zones or compartments. Through mass and energy balances in each compartment, the description of the temperature, humidity and coating mass concentration fields along the vertical coordinate was studied. Outlet air temperature values from the granulation of glass beads with distillate water in a pilot top-spray fluidized-bed granulator were used to validate the model. Hede et al. (2009) developed a model for a discontinuous top-spray fluidized-bed granulator, based on Ronsse et al.

(2007), to analyze how changing scale influences the process conditions from an up-scaling point of view. They found that the larger the pressure, the higher the droplets zone. Besides, the spray zone should ideally take up more than 50% of the fluidized-bed height.

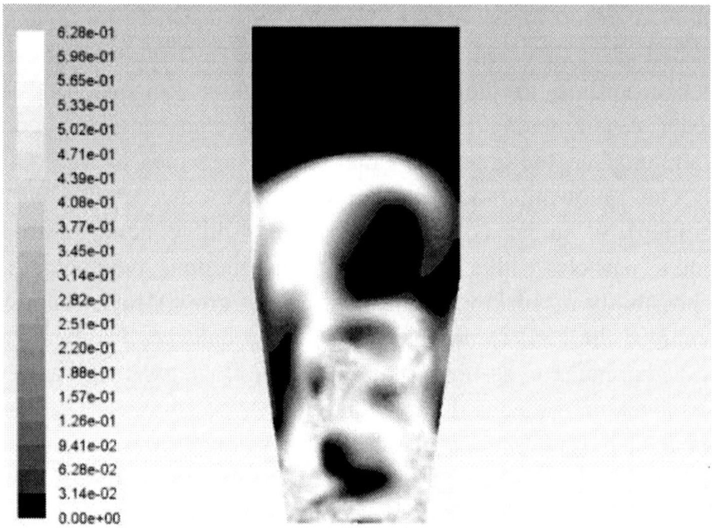

Figure 12. Volumetric fraction field contour map of urea particles in a fluidized-bed.

6.3.2. CFD Models

The mechanistic models have flow modeling restrictions that limit the accuracy in solving particles-fluid problems. With the evolution of computers, computational fluid dynamics (CFD) techniques emerged as an interesting alternative. Basically, the CFD approach divides the system volume into small cells (within each of which smoothly change of the variables can be assumed) and solves the mass, momentum (i.e., Navier-Stokes), heat and population balances within each cell to obtain velocity, concentration and temperature profiles, among other variables. The differential balance equations are discretized and solved iteratively until convergence. Figure 12 shows a contour map for the volumetric fraction field of small urea particles in a fluidized-bed (obtained with Ansys Fluent), where the chaotic nature of the fluidization can be visualized.

Both traditional models and those involved in CFD techniques belong to two-fluid modeling. In the two-fluid approach (also called Euler-Euler approach) both the fluid and particulate phases are considered as interpenetrating continua. Unlike traditional models, the CFD techniques conveniently compute the flow pattern and the momentum transfer solving the Navier-Stokes equation.

CFD is useful to study the granulator fluid dynamics. However, the dynamics of a particulate system cannot be described by the physical laws applied to solids, liquids and gases. This is because the interactions between particles are generally irreversible and extremely non linear. To solve the momentum equation for the particles as a continuum phase in the two fluid modeling, there are variables (such as the pressure and the strain stress) that must be quantified in terms of the solids flow. These properties depend on the particles velocity profiles, the volume fraction of each phase, the particles size, etc. The kinetic theory of granular flows (KTGF) is used in CFD to generate equations of closure for the internal momentum transport in the particulate phase (Goldschmidt et al., 2002). The KTGF is based on the kinetic theory of gases, applied on particulate flows taking non-conservative particle-particle collisions and the gas-particle drag into account. The particle velocity fluctuation plays an important role in the KTGF, producing an effective pressure in the particulate phase that must be taken into account in the particles momentum equation.

Rajniak et al. (2009) studied the wet agglomeration process in a bottom-spray Wurster granulator (i.e., a bottom-spray granulator where the fluidized-bed contains a draft tube to generate a circulating flow pattern) by using the CFD modeling based on the KTGF to evaluate the hydrodynamic behavior of the gas-solid mixture together with a PBE to quantify the particles growth. The results suggest that the particles mean size is practically the same in different locations of the fluidized-bed, supporting the suitability of the perfect mixing assumption in the Wurster granulator. Besides, by comparing the modeling results with experimental data, the authors determined the agglomeration kinetics. Chua et al. (2011a) applied CFD techniques to study the binder spreading time in the context of the fluidized-bed melt granulation process. In fact, they identified the melt droplets spreading time and the corresponding final spread area at equilibrium conditions to establish the circumstances at which the probability of size enlargement by agglomeration was higher.

The high computational cost and the nonlinearities present in the equations are such that CFD is not recommendable for complex systems. However, CFD allows reducing expensive experimental work and helps to assume the appropriate simplification hypothesis in the mechanistic models.

6.3.3. DEM Models

In recent years, understanding of the dynamic behavior of particulate systems has been achieved as a result of the rapid development of discrete element method (DEM) simulations (Zhu et al., 2008). Unlike the two-fluid approach, DEM solves the motion of each particle separately by means of the Newton law, considering the particle-particle, particle-fluid and particle-wall interaction forces. Hence, DEM simulations provide the trajectories of the individual particles and transient forces acting on them, which are extremely difficult (if not impossible) to obtain by physical experimentation (Zhu et al., 2008). DEM can be coupled with CFD techniques to describe the particulate and fluid phase respectively, facilitating the study of many particulate systems (Xu and Yu, 1997).

By simulation of a discontinuous fluidized-bed melt granulator in bubbling regime, considering particles size enlargement both by coating and agglomeration, Goldschmidt et al. (2003) found that low fluidization gas velocities increased the residence time of the particles in the spray zone, which caused bigger granules. Fries et al. (2011) presented a DEM model to explore the differences between a fluidized-bed top-spray granulator and a Wurster granulator. In the top-spray fluidized-bed granulator, the simulation results showed a wide residence time distribution of the particles in the spray zone causing an inhomogeneous granules wetting. In contrast, for the Wurster granulator the particles residence time in the spray zone was narrower and the wetting of the solids was more homogeneous as a result of the predefined particles flow pattern.

6.3.4. Multi-Scale Modeling

Fluidized-bed granulation is a complex process consisting of many interacting processes such as fluidization, atomization, drying/solidification, droplets adherence, collisions between particles and particulate mechanisms such as coating, agglomeration, attrition and nucleation. Therefore, to establish optimum granulation conditions trial-and-error procedures are often required (Ronsse et al., 2008).

DEM models can provide valuable information about the particle-particle interactions and the wetting kinetics. However, the particles number that DEM can handle (typically lower than 10^6) makes difficult their application to industrial-scale fluid beds (Goldschmidt et al., 2002) where the number of particles per chamber is typically three orders of magnitude higher (Bertin et al., 2010a). Two-fluid models through CFD techniques also provide information on the phases distribution and interaction, but require not only much computation time but also additional closure laws to describe all the interactions. On the other hand, mechanistic models are less computationally expensive although they provide less information about the process because they only capture the macroscopic behavior of the system. Thus, in order to link all the phenomena involved in granulation processes, a multi-scale approach combining the advantages of the different methods is needed.

In fact, nowadays there is a tendency to model particulate process by using a multi-scale modeling approach, (i.e., different levels of modeling) as the one presented in Figure 13 (van Sint Annaland et al., 2007). In each level, the phenomena that occur at that scale are analyzed. CFD techniques and DEM simulations can be used to evaluate the particle-particle and particle-fluid interactions, to calculate the velocity and concentration profiles and to develop closure equations useful to build large-scale models (by means of the mechanistic approach) consistent with the system physics.

Micro-scale phenomena such as the wetting kinetics and energy dissipation effects (related to the binder addition and mixing rates) directly influence the meso-scale particulate mechanisms as coating, agglomeration, breakage, attrition, nucleation (Poon et al., 2009). Moreover, these phenomena directly determine macroscopic properties such as the particles size (Figure 13).

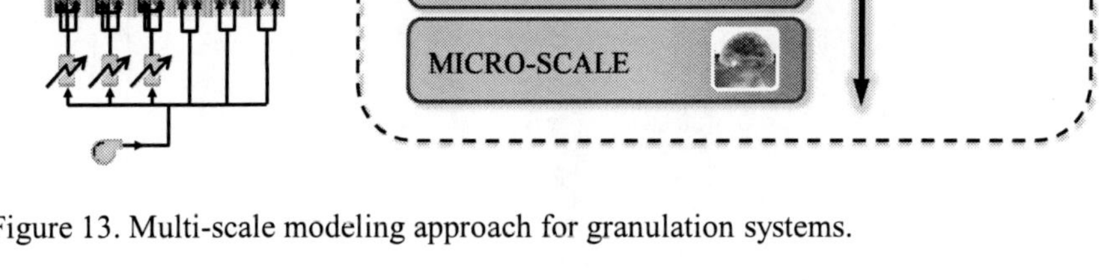

Figure 13. Multi-scale modeling approach for granulation systems.

Therefore, an integrated modeling structure including different scales is needed to combine models of different length and time scales in order to link the variables that control the granulation process.

6.4. Modeling of a Fluidized-Bed Urea Granulator

Despite the significant advances in modeling fluidized-bed granulators, the reported models are not applicable directly to the urea process. For this reason, Bertin et al. (2007, 2010a, 2010b and 2011) proposed different mathematical models to describe a large-scale urea fluidized-bed granulator, which consist of three growth chambers followed by three cooling ones. Bertin et al. (2007) represented the steady-state operation of the granulator by means of mass and energy balances which considered the behavior of the different phases that coexist within the unit (solid, liquid and gas) and the mass and energy exchanges between them. The results indicated that the urea dissolution heat is the most important thermal effect involved in the growth chambers; reinforcing the idea that melt granulation has a particular behavior that distinguishes it from the widespread studied wet granulation. The other main conclusions from the steady-state simulations can be summarized as: a) the urea solution is so concentrated and the mass and transfer rates are so high that water evaporation can be assumed complete, b) the gas and solids outlet temperatures are virtually the same (thus, a homogeneous model for the energy balance is enough to represent the process) and c) the chambers operate almost adiabatically.

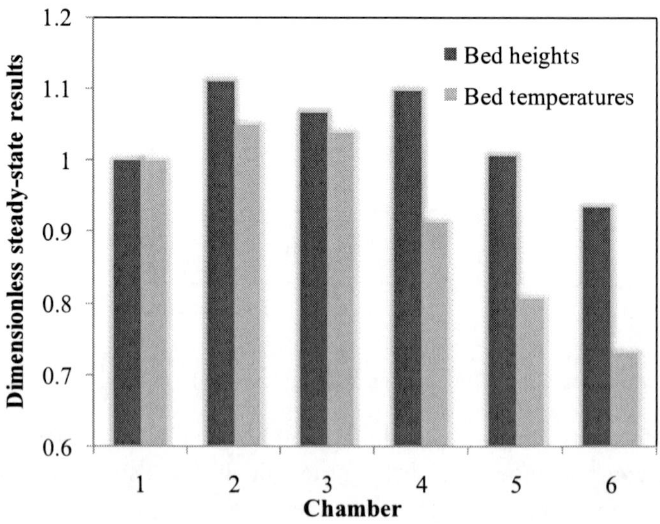

Figure 14. Fluidized-bed heights and mean temperatures obtained by steady-state simulation.

Figure 14 presents the steady-state simulation results for the fluidized-beds heights and average temperatures of each chamber, obtained at typical operating conditions and expressed as dimensionless variables with respect to the values of the first growth chamber. The fluidized-bed heights follow a trend that depends on the superficial air velocity profile and the population PSD within each chamber. Chambers 2 to 4 have the highest heights, being the difference between the highest and lowest values lower than 0.2 m. Figure 14 shows that the first chamber temperature is lower than that of the next one, basically because the seeds are recycled to the granulator at a much lower temperature. Downstream from this compartment, the bed temperature increases as a consequence of the addition of urea at a relatively high temperature (132 °C). Since the last three chambers are reserved for cooling purposes, the bed temperature decreases continuously from the third fluidized-bed toward the granulator outlet.

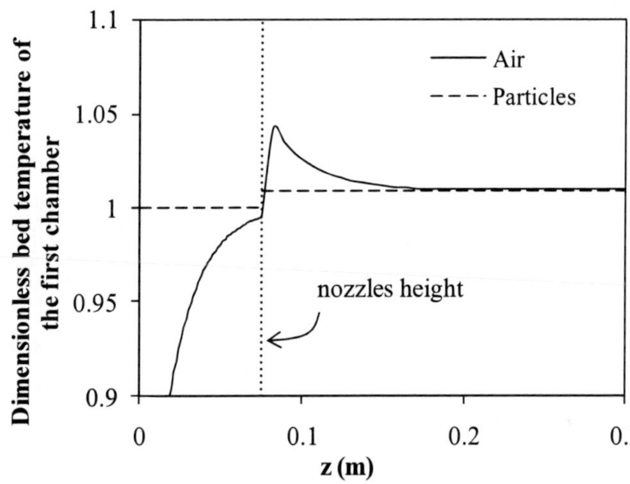

Figure 15. Steady-state axial temperature profiles (for the first 0.3 m of the total bed height) for the particles and air in the first growth chamber.

Figure 15 presents the steady-state axial temperature profiles for the particles and air in the first growth chamber, expressed as dimensionless variables with respect to the particles temperature at the granulator bottom (i.e., on the perforated plate air distributor). These results were obtained by the solution of a three-zone model: a lower zone which height is fixed at the atomizing nozzles height, a spray (or middle) zone which height depends on the spray jet penetration depth and an upper zone where only particles mixing takes place (Bertin et al., 2010b). Perfect mixing in each zone was assumed for the particles, while plug flow was considered for the air. The injection of the urea solution and atomization air at high temperature (i.e., close to the urea melting point) causes an increase of the bed temperature around the nozzles, which coupled with the dissolution heat released by the urea solidification, produces significant thermal gradients. Nevertheless, a single temperature value can be used to represent the thermal level in each fluidized-bed. In fact, the temperatures estimated by Bertin et al. (2007) assuming well-mixed flow pattern for the complete chamber (i.e. a multizone model was not used) are in very good agreement with industrial data, being the maximum deviations lower than 3 %.

Bertin et al. (2010a) formulated also an unsteady-state model based on mass, energy and momentum balances. This model was derived from the steady-state model (Bertin et al., 2007) incorporating simplifying hypotheses justified by the conclusions of the steady-state simulations. According to the high-capacity of the simulated industrial unit, most of the process variables (except chambers temperatures and porosities) exhibited slow dynamics, requiring in certain cases more than one hour to reach the final steady state. Based on a steady-state sensitivity analysis and the results given by some simple manual control actions, the product granulator discharge area and the fluidization air temperature and flowrate were identified as the more convenient manipulating variables to respectively control the fluidized-bed heights, temperatures and porosities in each chamber. Finally, Bertin et al. (2011) coupled the dynamic mass, energy and momentum balances with the population balance to predict the urea granulometry. Slow response times for the PSD of the granulator product were observed. It was also found that the seeds PSD is the variable that most affected the PSD of the stream leaving the granulator, being then important to make efforts to maintain the seeds product quality.

Figure 16 presents the SGN and UI values corresponding to the PSD of the seeds and granules that leave each chamber. The SGN of the granulator outlet stream gradually increases from the first to the third chamber and then remains constant. For the selected operating conditions, the granulator product SGN and UI increase from one chamber to another. The degree of mixing within the granulator has a significant effect on the granule size distribution. The PSD greatly broadens from the seeds PSD as a consequence of the exponential particles residence time distribution in each well-mixed fluidized growth chamber. However, Figure 16 shows an increase in the UI values through the growth chambers. This is because the UI is not a pure measure of the PSD spread. Although the particles residence time distribution contributes to decrease the UI (according to a higher dispersion), the granules growth tends to increase the aperture sizes that allow the passage of 5 and 90 wt% of the population and, thus, the UI values. According to the results of Figure 16, the later effect is dominant on the granulator UI trend, masking the increase in the PSD spread. The PSDs of chambers 4 to 6 (cooling compartments) are identical to that of chamber 3, as expected for steady-state and null growth rate. It is important to note that the SGN and UI values of the granulator product are not those of the final granular urea, because, as previously explained, the final product is obtained by screening of the granulator outlet solids stream.

The transient behavior of the large-scale fluidized-bed granulator shown low response times, indicating that the granulation unit would control the dynamics of the whole circuit. In effect, the transition times of the bed heights, pressure drops, solids mass holdups and solids mass flowrates were more than one hour, in agreement with the high capacity of the simulated industrial unit. On the other hand, the bed temperatures and porosities were less affected by changes in the inlet conditions, presenting faster dynamics. Considering the strong influence of the bed height, temperature, and porosity on the operation stability and product quality, the granulation unit should be carefully controlled.

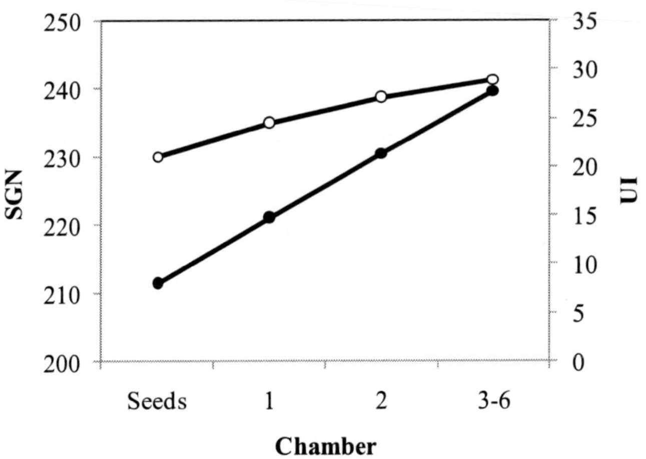

Figure 16. Particles SGN (•) and UI (○) along the large-scale fluidized-bed granulator.

Figures 17 and 18 present the evolution of the granulator product SGN and UI over time for +10% step changes in the seeds SGN, UI, flowrate and melt mass flowrate (performed 10 minutes after simulation from the initial steady-state). As it can be seen, the seeds PSD is the variable that most affected the outlet PSD. This result strongly suggests the importance of exploring different granulation circuit flowsheets aiming to keep the seeds granulometry close to the desired size distribution. Regarding the disturbances assayed in the seeds and melt flowrate, the 10% positive steps had opposite effects on the particles growth rate and, consequently, on the product SGN deviations.

The simulation tool developed for a large-scale urea fluidized-bed granulator is valuable to predict the process variables (flow, temperature, particle size distributions and pressure drops) and helps solving common operating problems in industrial units. In the studies mentioned in this section, only coating has been considered as the possible growth mechanism.

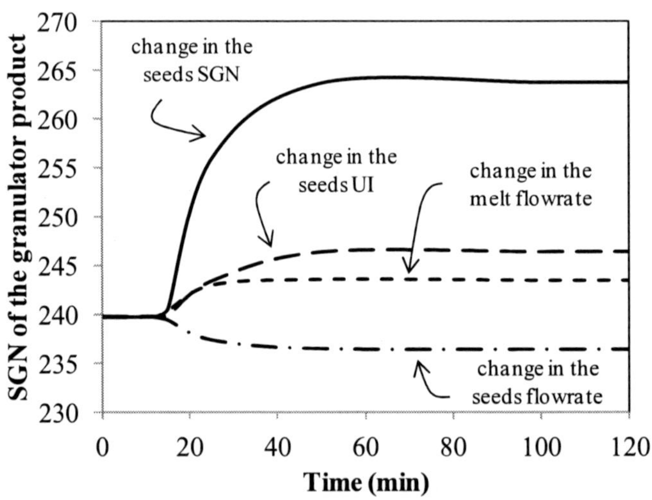

Figure 17. Granulator product SGN for a +10% change in the seeds SGN, UI, flowrate and melt mass flowrate, performed at 10 minutes of simulation.

The kinetics of this mechanism is directly proportional to the flowrate of urea melt fed to the system and the surface area of the particles in the granulator (Bertin et al., 2007). Although this size enlargement mechanism is the desired and most frequent under optimum operating conditions, deviations from these situations (such as a sudden decrease of the fluidization air flowrate) could favor particles agglomeration and potential defluidization. Therefore, it is clearly necessary to understand the influence of the operating variables on the competitive size enlargement processes (coating vs. agglomeration) that may occur in urea fluidized-bed granulators.

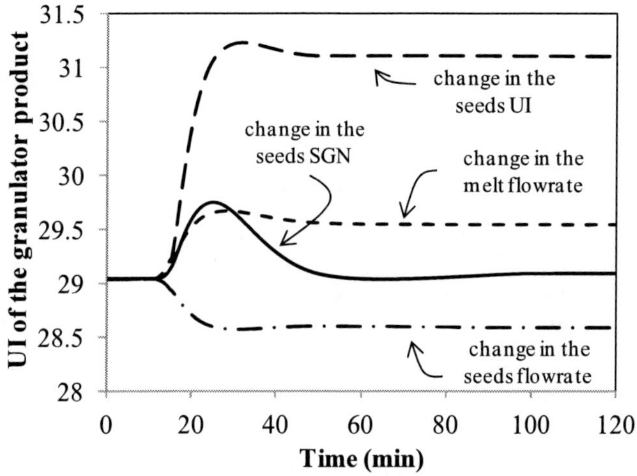

Figure 18. Granulator product UI for a +10% change in the seeds SGN, UI, flowrate and melt mass flowrate, performed at 10 minutes of simulation.

7. Effects of Operating Variables on Urea Growth Mechanisms

As above mentioned, granulation processes are usually classified according to the binder nature as wet, dry o melt. In the past, wet granulation was commonly used. Nowadays, dry or melt granulation are considered attractive strategies to overcome operating problems caused by moisture (Abberger et al., 2002).

The amount of articles related with wet granulation processes using aqueous solutions as liquid binders is vast. Among others, Smith and Nienow (1983), Pont et al. (2001) and Hemati et al. (2003), have studied the influence of the process variables and physicochemical properties on the particles growth kinetics when aqueous solutions are atomized into beds of solids fluidized by hot air. These experimental studies revealed that many variables can affect the particles growth mechanism: binder (composition, viscosity, surface tension, flowrate, drop size), seeds properties (size, shape, porosity), atomization and fluidization air flowrates, nozzle location, bed temperature, etc. Among others, the dependence of the growth regime (agglomeration or coating) and growth rate upon the excess gas velocity (i.e., difference between the actual fluidization velocity and the minimum fluidization one) was empirically established. For wet granulation and low excess gas velocities, agglomeration was identified as the main particle growth mechanism. For higher gas velocities the particles circulation rate increased, improving the liquid distribution on the seeds surface and reducing the bed quenching by formation of lumps (i.e., large agglomerates). Moreover, higher velocities increased the frequency and energy of the inter-particle collisions and particle-wall impacts responsible for the breakage of the binder solid bridge that may have been formed between primary seeds (Smith and Nienow, 1983; Hemati et al., 2003). The same behavior was observed, through experimental work, for different granulation systems (i.e., operating variables and material parameters). Besides, the growth regimes were found to be very sensitive to the product and type of granulation unit. In order to generalize the findings of many authors about wet granulation, a mayor turning point was the paper published by Ennis et al. (1991), who proposed a physical-based model for predicting the growth behavior of granules. Even though that model was simple and based on many assumptions, it is still helpful to predict the growth regimes using measurable variables.

The understanding of the mechanisms prevailing in the granulation process is a prerequisite for obtaining proper control over powder properties (Boerefijn and Hounslow, 2005). Unfortunately, the theories developed for wet granulation are not fully appropriate in describing fluidized-bed melt granulation (Walker et al., 2005; Tan et al., 2006). Since the use of molten binder eliminates the need of solvents, the fluidized hot melt granulation has received considerable attention in recent years. As for wet granulation, many authors were focused on revealing the influence of some of the most important experimental variables on the product quality. The studies can be grouped into two categories: 1) the binder and the seeds are initially loaded into the bed as solid powders and are then fluidized by means of hot air and 2) the granulation is performed by atomizing a molten binder within a bed fluidized by relatively cool air. Within the first group, among others, Zhai et al. (2009), Walker et al. (2005) and Walker et al. (2006) explore the effect of the granulation time, binder content and viscosity, and seeds size on the granules final size and growth mechanisms. With respect to the second group, Abberger et al. (2002), Seo et al. (2002), Tan et al. (2006) and Boerefijn and Hounslow (2005) studied the effects of binder spray rate and droplet size, seeds size, bed temperature, atomization air pressure and fluidization air velocity on the performance of fluidized-bed melt granulation. These studies used polyethylene glycol (PEG) as a model binder and glass ballotini or lactose as seeds. In addition, the binder flowrate was generally well below 200 g/min, the seeds sizes were lower than 200 μm and the ratios of total binder atomized to the initial bed mass were quite low. Even though the conclusions of these studies provide valuable insights into the melt granulation field, they cannot be applied straightforward to the production of urea granules.

Besides using a very concentrated urea solution (95-97 wt%) as binder, the urea granulation is carried out with seeds of about 2 mm, droplet sizes around 20 μm, high urea melt to seeds mass ratios (about 50%), high urea melt flowrates and short granulation times. Therefore, although lot of effort has been done in order to understand the basic principles that govern the melt granulation, there is still need to perform experimental studies using the materials of interest and appropriate operating conditions to mimic the phenomena that occur in the industrial practice.

Regarding urea granulation, Roy et al. (2010) performed experiments to explore criteria to identify the endpoint of batch fluidized-bed granulation and to analyze the effect of the operating variables on the granules size using urea powder as a model system. The authors selected urea granules between 0.3 to 0.5 mm as initial particle sizes and saturated solutions with urea contents within the range 73-91 wt% as binder. The granulation runs were carried out in a top-spray lab-scale fluidized-bed granulator. The variables chosen by the authors were not entirely representative of the industrial process. For this reason our research team performed several urea-granulation experiments in a pilot-scale fluidized-bed unit taking into account the main features of the industrial process, being the main objective to identify the granules growth mechanism regimes in order to avoid those operating conditions that could give granules with poor properties.

Tan et al. (2006) and Chua et al. (2011b) reported possible growth mechanisms that can occur during fluidized-bed hot melt granulation. Based on the acknowledged feasible growth mechanisms and our own experimental evidences, Figure 19 shows a schematic of the microscopic events that can occur during the urea granulation (considering that urea is sprayed at around its melting temperature and the bed is maintained at approximately 100 °C). It is worth to remark that the urea melt droplets size is about 100 times smaller than the seeds size; therefore Figure 19 is not representative of the true particles scale. Some of the tiny droplets may be not captured by the seeds and, thus, be rapidly converted to fine powder (event 1), which can be entrained or recaptured by the solids bed. Event 2 corresponds to the droplet-particle collision; as it can be seen, the liquid binder coverage can either solidify before any other granule touches the wet one (event 3) or be in contact with another particle (event 4). The agglomeration can be successful (event 5) or, depending on the rate of the binder solidification, the liquid bridge can be broken (event 6). Finally, and according to the fluidized bed turbulence, agglomerated particles can be detached through event 7. For the production of urea granules, path 3 is clearly the desired growth mechanism. In order to avoid an exponential granules growth (agglomeration, i.e., scenario 5 of Figure 19) and an eventual bed defluidization (known as bed quenching), the operating variables have to be selected carefully. In this context, the aim of this section is to reveal the influence of some operating conditions on the growth mechanisms and on the most important urea granules properties.

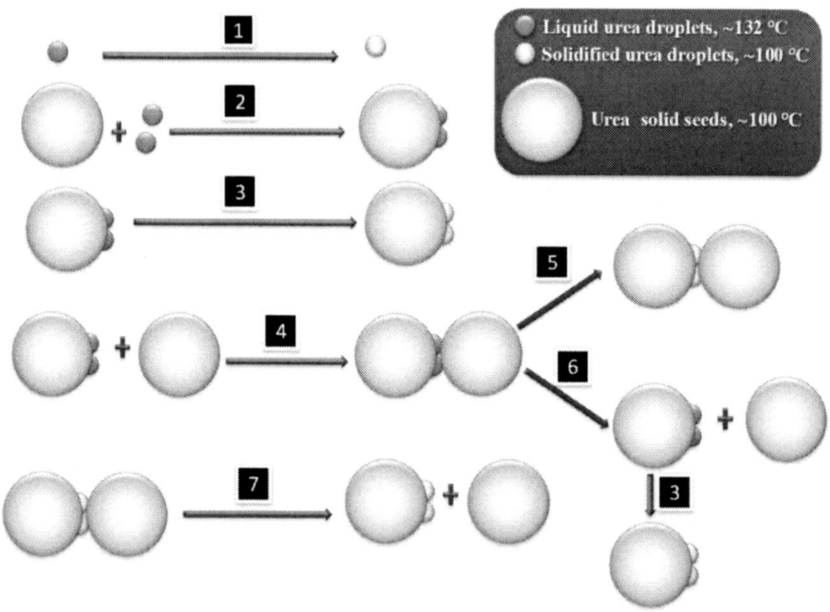

Figure 19. Schematic of probable microscopic events occurring during urea melt granulation.

7.1. Experimental Set-Up and Procedure

A schematic diagram of the experimental device is shown in Figure 20. The experiments were carried out in a batch fluidized-bed granulator constituted by a stainless steel bottom conical vessel (1), which is inclined 6° with respect to the vertical axis (see Table 4), and a cylindrical column (6) on top of it. The air distributor is a stainless steel perforated plate (2) of 3 mm thick with a porosity of 6% (156 holes, 3 mm each hole diameter). To prevent solid particles from flowing back into the space below the distributor, an ASTM #30 mesh stainless steel screen covers the air distributor plate. The fluidizing air was supplied by a centrifugal blower (3). Before entering the bed, the fluidizing air flow rate was measured by an orifice flow-meter (4) and preheated by an electrical heater (9 kW) (5) to maintain the bed temperature at the desired level. The bed and grid pressure drops were continuously measured by differential pressure instruments, or alternatively manually by water U-tube manometers. The elutriated fine particles were collected by a set of three filter bags located at the top of the fluidized-bed freeboard (6). These filters were periodically blown back by air pulses to disengage the particulate matter. The feed solution (urea melt) was prepared in an oil-heated tank (7) by typically adding 1 kg of urea, a given water volume to reach the desired urea concentration and a tiny amount of food dye to easily monitor the fluidized-bed granulation through the unit observation window. The urea melt tank was placed on a scale and kept at constant temperature (≈130 °C). The urea concentrated solution was delivered to an internal mixing two-fluid spray nozzle (8) located just above the air distributor by means of a given compressed air flowrate, which was preheated up to 130 °C in the oil reservoir before entering into de urea solution tank. The atomizing air flowrate was controlled by a valve and measured by a rotameter. The external tube-skin temperature of the urea-solution line (from the hot container to the spray nozzle) was controlled through an electric heat tracing system. A Programmable Logic Control system (PLC) was used to register and control several process variables.

For each run, a batch size of approximately 2 kg of urea particles of a given cut size (seeds) was initially charged into the bed chamber. The seeds were fluidized with hot inlet air until the desired bed temperature level was achieved.

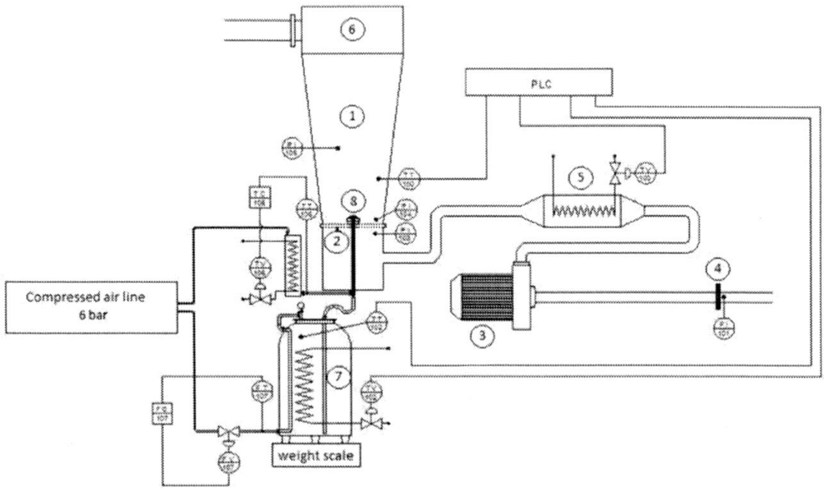

Figure 20. Schematic representation of the experimental set-up.

Table 4. Fluidized bed granulator geometrical parameters

Parameter	Value
Conical section height, H_{CO} (m)	0.7
Cylindrical section height, H_{Cy} (m)	0.45
Bottom diameter, D_{bot} (m)	0.15
Top diameter, D_{top} (m)	0.3
Angle γ (°)	6

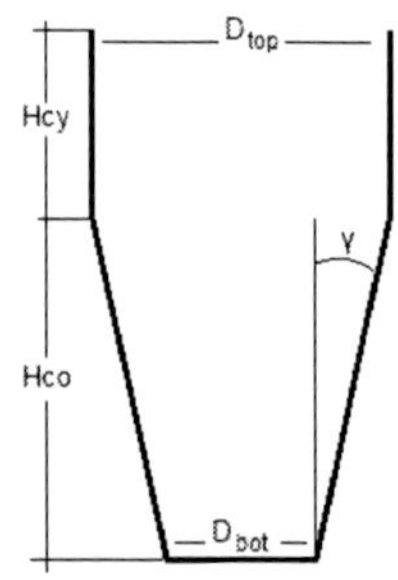

Table 5. Experimental conditions for the fluidization air velocity parametric study

Variable	Value
Seeds diameter (between ASTM sieves #8 and #10), mm	2.38-2.8
Seeds arithmetic mass mean diameter, mm	2.59
Seeds minimum fluidization air velocity*, m/min	49.2
Superficial fluidization air velocity*, m/min	132-270
Atomization air flowrate, m^3/min	0.03
Urea melt flowrate, kg/min	0.276 ± 5.6%
Urea melt concentration, wt%	97.5 ± 0.5%
Bed temperature, °C	100
Granulation time, min	3.5-3.9

(*) measured at the bed entrance and expressed at atmospheric pressure and 20 °C.

Then, the urea solution was sprayed at constant mass flowrate. This flowrate was computed from the linear time profile of the recorded urea-solution container weight. Once the binder spraying was stopped, the granules were immediately cooled down using air at room temperature. The collected granular product was set aside for further characterization. The powder deposited onto the granulator walls, the fines collected in the filter bags, and the urea solution that remained in the oil-heated tank after each experiment were weighted for mass balance closure calculations. The final urea solution was withdrawn by using a pipette, the material was stored in a flask hermetically sealed and, as soon as it solidified, it was subjected to a moisture content assay to establish the final urea solution concentration. Several experiments were performed to understand the effects of the operating variables on the growth mechanisms, granulation efficiency and granules properties. To this end, the superficial fluidization air velocity at the bed entrance was varied from 132 to 270 m/min, the arithmetic mass mean diameter of the urea seeds ranged from 1.5 to 3.7 mm, the urea melt flowrate from 0.24 to 1.02 kg/min, the bed temperature from 90 to 110 °C, the atomization air from 0.024 to 0.036 m^3/min and the urea solution concentrations between 89 and 97 wt%. Since the fluidization air velocity is a key operating parameter (Tan etal., 2006), the results regarding the influence of this variable on the granulation performance are presented in this chapter. Table 5 summarizes the experimental conditions selected for that parametric study.

7.2. Granular Product Characterization

Particles size distribution. Due to the short granulation times, the particle size distribution (PSD) was evaluated for the solids collected at the end of the experiments.

A riffle splitter was used to accurately divide the total sample (about 3 kg) into representative samples. A batch of 0.8 kg was sieved in a vibratory sieve shaker (Zonytest, Argentina). A series of 9 ASTM standard sieves (# 4, 5, 6, 7, 8, 10, 12, 14 and 16) was employed. The sample was shaken for 20 min at about 2400 strokes per minute.

Granulation efficiency (η). This efficiency, which was defined as the ratio between the mass of sprayed urea collected in the granular product and the total mass of urea sprayed into the bed during the whole experiment, can be expressed as:

$$\eta = \frac{M_{uf} - M_{u0}}{\dot{M}_{melt} X_u t} 100 \qquad (4)$$

where and are the final and initial mass of urea particles in the bed, represents the urea solution flowrate, while X_u and t are the urea concentration of the sprayed solution and the spraying time, respectively. does not include neither the mass retained on the unit walls nor in the filter bags.

Arithmetic mass mean diameter (d_m). This diameter was determined from the sieve analysis data as:

$$d_m = \frac{\sum w_i \overline{dp}_i}{\sum w_i} \qquad (5)$$

where $\overline{dp}_i$ is the arithmetic mean diameter of size class (average of the apertures of two consecutive sieves) and is the mass fraction (wt%) collected by sieve analysis within the ith size interval.

Size Guide Number (SGN). As defined in Section 2, SGN represents the particle size in millimeters for which 50 % by weight of the product is coarser and 50 % is finer multiplied by 100 (i.e., the SGN is 100 times the mass median of the population).

Uniformity index (UI). As also defined in Section 2, this index characterizes the spread of the product particle size distribution (PSD) and is defined as the ratio of the opening size that would let pass 5 wt% of the corresponding sample to the opening size that would let pass 90 wt%, multiplied by 100.

Agglomerates. The agglomerates are particles constituted by two or more primary seeds. Considering the initial granules size and the selected series of sieves, the particles collected in the granular product bigger than 4 mm were defined as agglomerated material.

Fines. The particles smaller than 2.38 mm (i.e., the minimum initial particles diameter) were labeled as fines. Laser diffraction analysis of the collected fines revealed that their sizes were between 32-75 μm; i.e. very fine powder. In this study three different types of fines were considered:

Granular-product fines: Fines in the collected granular product.

Wall fines: Powder deposited onto the granulator walls. Since the fluidized-bed height was lower than 0.2 m (it varies with the fluidization velocity), the fluidized-bed walls surface represented a very small fraction of the total granulator walls surface. Therefore, these fines were mainly associated to particles elutriated from the bed.

Filter bags fines: Powder retained in the filters.

Granule moisture content. Granules samples of 10 g (immediately collected after each granulation run) were dried at 95 °C for 9 min in an Ohaus MB45 moisture analyzer. This treatment profile was defined, using some selected samples, to reproduce the moisture content determined by the Karl-Fisher method. The moisture content was determined by triplicate.

Crushing Strength. A testing machine Instron model 3369 in compression mode at a speed of 2 mm/min was used. For each experiment, twenty crushing strength measurements of granules that blind the apertures of an ASTM #7 sieve were performed, being the reported value the average of those results.

Particle morphology. Some selected granules and their cuts were assessed in an EVO 40-XVP, LEO Scanning Electron Microscope (SEM). Previously, the samples were metalized with gold in a PELCO 91000 sputter coater.

7.3. Results and Discussion

Figure 21 presents the mass fractions of agglomerates within the granulator product and granular-product fines with respect to the total mass of sprayed urea, as a function of the fluidization air velocity at the bed entrance. Two operating regions were clearly distinguished:

- $2.7v_{mf}$ (132 m/min) $\leq \leq$ (180 m/min): The particles enlargement was due to both coating and agglomeration, being more important the coalescence as the fluidization velocity decreases from the critical value of 180 m/min.
- $3.7v_{mf}$ (180 m/min) $\leq \leq$ (270 m/min): For relatively high fluidization air velocities, the growth by coating was the dominant mechanism, being the agglomerates mass fractions lower than 1.5 wt%.

For very low fluidization air velocities, either the bed quenched or very large agglomerates (as the lump presented in the picture at the top of Figure 21) were produced.

For velocities higher than $3.7v_{mf}$, a free flowing material was obtained (see picture displayed at the bottom of Figure 21). This result is in good agreement with the findings reported for either wet granulation (Smith and Nienow, 1983) or melt granulation (Tan et al., 2006). For these two different types of granulation, faster powder circulation through the spray zone (given by higher fluidization air velocities) allowed improving the binder distribution and reducing the chances of bed quenching or the formation of large agglomerates. Under relatively high fluidization air velocities, the disruptive forces (basically determined by the intense bed turbulence), were higher than the liquid or solid bridges binding forces. Consequently, coating appeared as the dominant growth path (events 3 and 7 of Figure 19). Even though, it is well known that the fluidization air velocity is a key parameter in controlling the final granules attributes, the value of the critical fluidization velocity needs to be predicted for each granulation system (seeds-binder-apparatus).

Figure 21 shows that the mass fraction of fines particles collected in the granular product, with respect to the total mass of sprayed urea, was lower than 1.4 wt% for all the studied fluidization air velocities. The origin of these fines could be either by nucleation (i.e., solidified binder droplets) or agglomerates attrition (which are very unstable and poor consolidated particles).

Although it is low, the highest granular-product fines mass fraction was obtained for the lowest fluidization air velocity tested. According to the expected terminal velocities, these fines could have been elutriated. However and it is can be seen in Figure 22, which presents the granulation efficiency and the mass fraction of fines (with respect to the total mass of sprayed urea) collected on the unit walls and in the filter bags as a function of the fluidization air velocity, the amount of fines entrained was minimum. At this operating condition, the bottom portion of the bed quenched almost instantaneously by fast agglomeration around the nozzle, preventing the normal delivery of the binder droplets. The partial defluidization and poor agitation caused by lumps formation could be responsible for the entrapment of fines within the bed. On the other hand, the lowest granular-product fines mass fraction was found for the minimum velocity that ensured particles growth by coating. In between, as the fluidization air velocity was increased, lower amounts of granular-product fines and higher quantities of unit walls and filter bags fines were obtained. These trends could be explained in terms of the improvement in fluidization quality (from a partially fluidized bed to a bubbling one, given by the higher air flowrates).

For gas fluidization velocities higher than 180 m/min, the bed was totally fluidized and the bed expansion increased with the air flowrate. As a result, the distance from the spray zone height to the bed surface increased, favoring further fines capture by other particles before reaching the freeboard. This could be the reason for the increase of fines in the granular-product (Figure 21) and the decrease in the unit walls and filter bags (Figure 22). Particularly, for gas velocities higher than approximately 240 m/min, the amount of fines on the unit walls and in the filter was less sensitive to changes in the air flowrate, indicating a balance between the elutriation rate and the bed expansion.

The lost in granulation efficiency shown in Figure 22 for gas fluidization velocities lower than 180 m/min was basically caused by the increase in the mass of fines collected on the walls and in the filter bags. For this reason, both curves are symmetrical. As above pointed out, almost all the atomized urea-melt flowrate solidified onto the relative big agglomerates rapidly formed at low fluidization velocities and located over the nozzle. Therefore, nearly all the sprayed urea was collected within the material discharged from the granulation unit. Even though very high granulation efficiencies were found, this operating condition should be completely avoided.

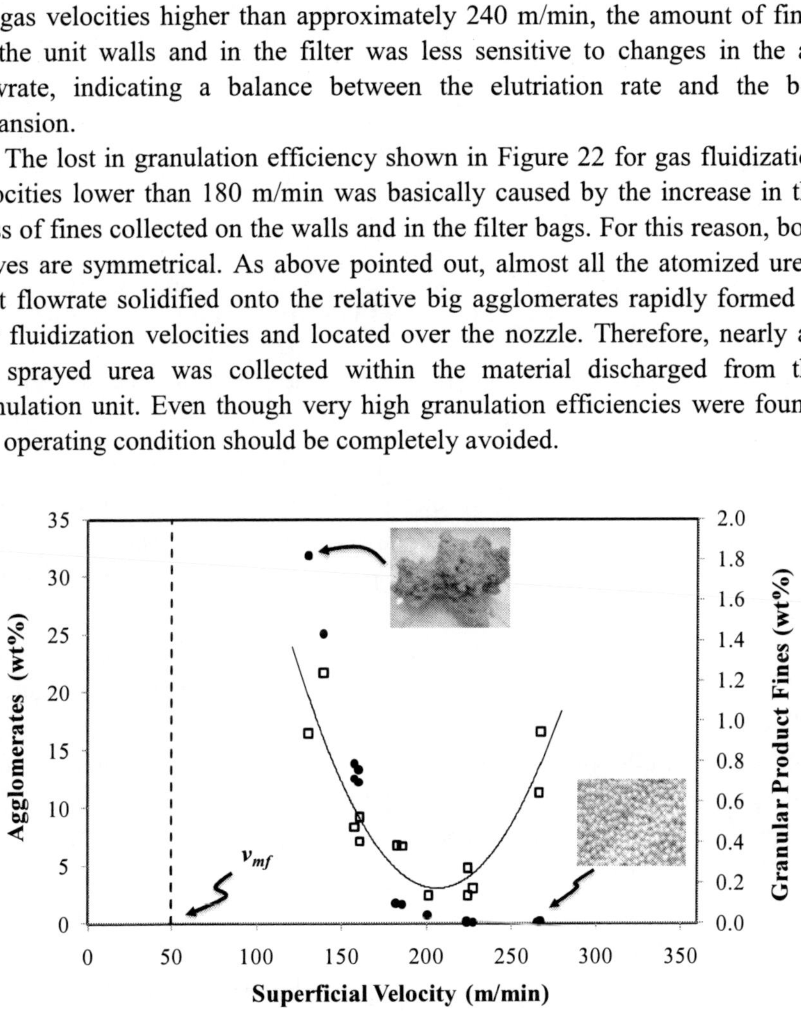

Figure 21. Effect of the fluidization gas velocity on the growth regime. (●) Agglomerates; (□) Granular-Product Fines. Curves are trend lines.

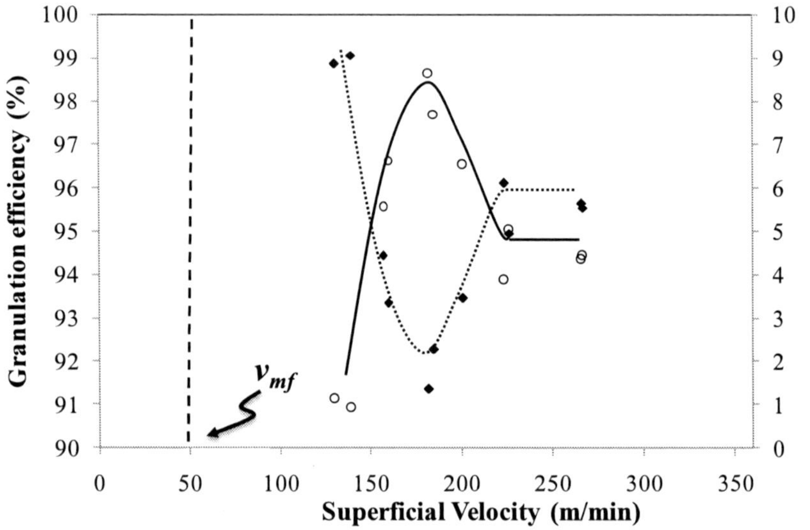

Figure 22. Effect of the fluidization gas velocity on the granulation efficiency (♦) and wall and filters fines (○). Curves are trend lines.

The granule moisture content as well as the crushing strength as a function of the fluidization air velocity are shown in Figure 23. For high gas velocities, the intense heat and mass transfer allowed reducing the particles moisture content. The marketable urea granules require moisture contents lower than 0.20 wt%, therefore velocities higher than 180 m/min were needed to achieve the adequate granules dehydration. Regarding the crushing strength, values higher than 3 kg_f are recommended for the urea bulk handling. Figure 23 indicates that the higher the air velocities (intense particles rebound and granules-unit walls collisions) the higher the granules consolidation and, thus, the crushing strength. Velocities higher than 180 m/min gave granules of 2.8 mm (ASTM#7) hard enough and with very low agglomerates mass fractions.

Figure 24 presents the population mass median (SGN/100), the arithmetic mass mean diameter and the UI obtained for granulations performed at different fluidization velocities. The mass mean diameter decreased significantly with increases in the air velocity as a consequence of the agglomerates disappearance. On the other hand, the median did not exhibit important changes as the fluidization velocity was modified. For gas velocities higher than 180 m/min (i.e., mass fraction of aggregates negligible), the median and mass mean became similar. This fact indicates that the PSDs tend to be normal mass distributions. Figure 24 also indicates that the final PSDs became narrower as the fluidization velocity was increased. As pointed out by other systems (Tan et al., 2006), the higher the air flowrate the better the binder distribution and, consequently, the lower the distribution spread (i.e., higher UI).

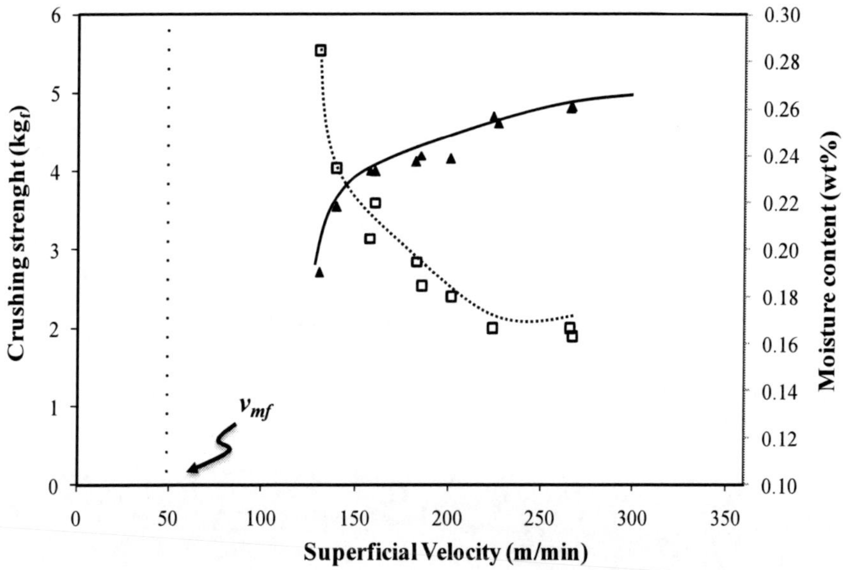

Figure 23. Effect of the fluidization gas velocity on crushing strength (▲) and moisture content (□). Curves are trend lines.

The air velocity also affected the morphology of the particles. Figures 25.a and 25.b show the SEM micrographs of granular product collected at 132 and at 270 m/min, respectively.

Particles with more homogeneous surfaces were obtained at higher gas velocities, as a consequence of the more vigorous particle-particle and particle-walls collisions. This observation is consistent with the higher crushing strengths found at higher air velocities.

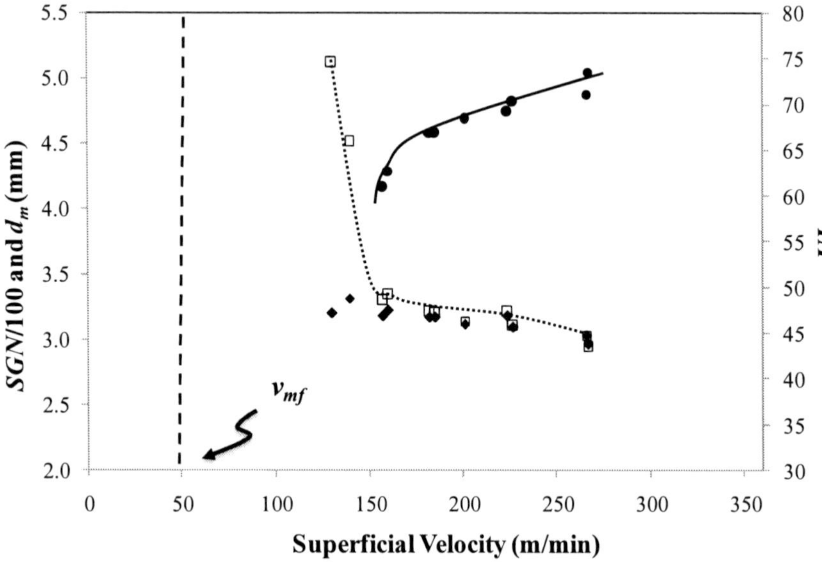

Figure 24. Effect of the fluidization gas velocity on SGN (♦), d_m (□) and UI (●). Curves are trend lines.

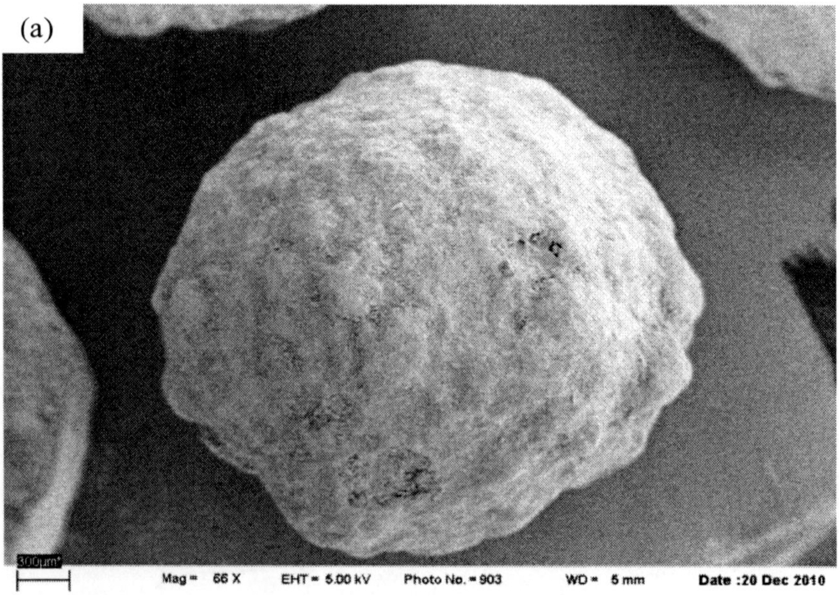

Figure 25.a. SEM micrograph of granules obtained at fluidization air velocitiy 132 m/min.

Figure 25. b. SEM micrograph of granules obtained at fluidization air velocity of 270 m/min.

For a ratio of atomized urea mass to initial bed mass of about 50%, binder urea concentrations around 96 wt%, bed temperature of approximately 100 °C and initial seeds of around 2 mm (typical industrial variables), fluidization air velocities above almost 4 times the seeds minimum fluidization one were required to favor the particles size enlargement by coating (the preferred growth mechanism to produce urea granules with adequate attributes to be commercialized as solid fertilizer). Velocities around also provided granules with low moisture content, adequate crushing strength, narrow PSD and rounded particles. The obtained results revealed the need to developed tools, based on experimental evidence, to predict the operating conditions of the urea fluidized-bed granulation that lead to undesired growth mechanisms. In this sense, further work is being done in order to build a regime map to predict the process variables that guarantee the production of marketable urea granules.

8. Modeling and Simulation of Fluidized-Bed Granulation Circuit

In this section, a detailed description of the peripheral units constituting the granulation circuit of a high capacity urea plant is given. Furthermore, studies regarding the steady-state and dynamic performance of the system are presented. Finally, and with the aim of showing the applications and capabilities of the developed dynamic circuit simulator, an optimal control problem for increasing plant throughput is discussed as example.

8.1. Peripheral Devices

As previously mentioned, even though the granulator is the core circuit unit, other devices are required to produce particles in the commercial size range and to generate the seeds that are fed back to the granulator as recycle stream (Figure 7). As the fluidized-bed granulator has already been introduced, this section is focused on the remaining circuit units: the double-roll crusher, the double-deck vibrating screen and the fluidized-bed cooler.

8.1.1. Double-Roll Crusher

The particles size reduction is considered a critical operation within the granulation process since it defines, together with the screen classification, the PSD of the seeds fed to the granulator which affects the final product quality. Several authors have investigated the incidence of the crusher performance on the seeds PSD. It has been reported that the crushing of the screen oversize material has a decisive influence on the circuit stability. In fact, variations in the average diameter of the stream that leaves the crusher may cause the instability of the system (Heinrich et al., 2002; Heinrich et al., 2003; Drechsler et al., 2005; Dosta et al., 2010). Furthermore, the size reduction is considered as a highly inefficient operation. In fact, only 5% of the energy is used for particle breakage and the unit design and scale-up are directly associated with the manufacturer experience (Rhodes, 1998).

There are many size reduction equipments that find applicability in a wide variety of industries, especially in the mineral processing one. Among others, the ball, hammer and autogenous mills as well as the jaw, gyratory and roll crushers can be mentioned. Usually, double-roll crushers are used in urea granulation circuits. These devices are constituted by two pairs of rolls (arranged in series) that rotate in opposite directions at different speeds. The rolls can be smooth, corrugated or toothed, being the distance between them (known as GAP) a key variable parameter. In the breakage of urea, the double-roll crusher is preferred over other comminuting equipments because narrow size distributions, low dust and limited noise generation are expected (Campbell and Webb, 2001). For the double-roll crusher two parameters can be modified: the upper and the lower gaps.

The ball and hammer mills have been extensively studied because of their applicability in the well-known mineral processing industry. Many mathematical models have been developed to accurately describe their performances (Adetayo et al., 1995; Austin, 1971/1972; Austin and Bhatia, 1971/1972; Austin and Cho, 2002; Benzer et al., 2001; Kis et al., 2006). As the double-roll crusher has less applicability in the mineral plants, few researchers have studied its modeling. It is important to note that the well-known ball mill model cannot be directly used for roll crushers. In fact, the double-roll crusher operates in a different way than reservoir type of mills and, therefore, must receive a different treatment. In ball mills the particles have a residence time during which they suffer repeatedly breaks. Breakage in double-roll crushers ('once-through' equipment) occurs instantaneously while the material is pulled into the rolls; thus, all the particles pass through the set gap (Austin, 1971/1972).

The crusher model involved in the urea granulation circuit simulator, developed by our research group, has been reported in a previous contribution (Cotabarren et al., 2008) and is based on that proposed by Austin and coworkers (Austin et al., 1980) for the mineral processing industry. It allows predicting the performance of the crusher (i.e., the outlet product PSD) as a function of the feed PSD (screens oversize fraction) and the gaps settings of both pairs of rolls. The parameters corresponding to this model were fitted to properly describe available crusher data from a large-scale urea granulation plant (Cotabarren et al., 2008).

8.1.2. Double-Deck Vibrating Screens

Screening is probably the oldest and most widely-used physical size separation method in industrial operations for continuous classification of solid streams. The employment of screens has spread to a variety of engineering categories, from the traditional mining sector to the contemporary fast-growing food and pharmaceutical engineering (Li et al., 2003). The fertilizer industry requires, as aforementioned, a classification step that is usually performed by double-deck vibrating screens.

Vibrating screens have been extensively studied by numerous authors in the context of the mining processing industry (Karra, 1979; Whiten, 1972; Ferrara et al., 1988; Subasinghe et al., 1989; Subasinghe et al., 1990; Soldinger, 1999). Models in the literature can be classified as phenomenological, empirical and numerical; being based on the theory of the screening process, empirical data and computer solutions of Newtonian mechanics, respectively (Wills and Napier-Munn, 2006).

Within the phenomenological models two different approaches, the kinetic and probabilistic, have been used to represent these screening operations. The probabilistic approach is based on the probability of a particle passing through the aperture of the screen (Whiten, 1972; Subasinghe et al., 1989). On the other hand, the kinetic approach defines the screening performance as a rate process that varies with the distance along the screen and depends on the amount and PSD of the material being processed. According to Ferrara et al. (1988) a zero-order process occurs at the beginning, while as the quantity of material on the screen declines, a first-order process takes place. Subasinghe et al. (1990) described the screening operation by an alternative approach, which uses two first-order rate processes. Soldinger (1999) established that the screening process involves two mechanisms, stratification and passage of particles through the screen apertures. The rate of stratification varies with the proportion of fine material and the particle sizes, while the rate of passage depends on the probability that the particles will pass through the apertures as well as the amount of fine material on the screen surface. This last mechanism could be considered as a combination of the kinetic and probabilistic approaches.

The empirical models aim to predict the quantity of undersize that can pass through the screen based on a theoretical capacity (Wills and Napier-Munn, 2006). This base capacity is affected by a set of correction factors which account for, among others, the effect of oversize, half-size, and near-size material. Other correction factors include whether the screen is a top or lower deck on a multi-deck unit, type of aperture, material density, etc. Due to their simplicity, the empirical models are preferred for the simulation of industrial screens. In fact, many commercial simulators such as Aspen Plus (2005), Modsim (2008) and Moly-Cop Tools (2008) have empirical models implemented in their routines.

The screen model included in the developed urea granulation circuit simulator was reported by Cotabarren et al. (2009). It is based on the empirical model proposed by Karra (1979) that represents the classification operation by determining the oversize partition coefficients. The screen mathematical model given by Karra (1979) was found, after thorough discrimination of several empirical, probabilistic and kinetic models, as the most suitable one to reproduce industrial data (Cotabarren et al. 2009).

8.1.3. Fluidized-Bed Cooler

As it was above described, once the particles leave the granulation unit, the product stream is sent to a fluidized-bed cooler located on a lower plant level by means of gravity (see Figure 7). The purpose of this unit is to decrease the urea granules temperature in order to avoid product lumps formation during further handling and bulk storage.

The fluidized-bed cooler is analogous to each cooling granulator chamber. Indeed, the particles that enter in this unit are fluidized and cooled by means of ambient air that is provided by a blower. The fluidized-bed cooler mathematical model has been first reported by Cotabarren et al. (2010) and is mainly constituted by the same dynamic mass, energy and population balance equations that describe the continuous well stirred granulator cooling chambers.

8.1.4. Steady-State and Dynamics of the Granulation Circuit

It is well known that in fertilizer granulation plants only a relatively small fraction of the material leaving the granulator is within the specified product size range; therefore, high recycle ratios are common. The characteristics of the recycle, which are the consequence of what happened previously in the granulator, influence what will happen later on in that unit. Thus, cycling surging and drifting of particles could take place. In extreme cases, these periodical oscillations coupled with large dead times can result in plant shut down or permanent variations in the plant capacity as well as product quality. To minimize these problems, it is necessary to have a fundamental understanding of the effects of the recycled material on the behavior of the granulation circuit (Adetayo et al., 1995; Heinrich et al., 2003). In view of this, the relatively high current installed world urea capacity and its forecasted expansion, the modeling and simulation to optimize the urea granulation circuits operation will play an important role in this fertilizer economy.

Little information is available in the open literature on urea granulation circuits. Wildeboer (1998), Adetayo (1993), Adetayo et al. (1995), Zhang et al. (2000), Balliu (2005) and Balliu and Cameron (2007) studied the dynamics and stability of drum wet granulation circuits by solving mass, energy and population balances. Due to the wet nature of the processes and the type of granulators within those circuits, the results of the sensitivity analysis carried out by these authors are not valid for the urea melt granulation that takes place in multichamber fluidized-bed units. Although Heinrich et al. (2002), Heinrich et al. (2003), Drechsler et al. (2005) and Radichkov et al. (2006) studied circuits including fluidized-bed granulators, all those continuous units were constituted by just one chamber where wet granulation processes occurred. In addition, either these authors assumed constant mass holdup for the granulator and/or hypothetical PSDs for the outlet crusher stream (i.e., the crusher operation was not modeled). Therefore, those works about circuit dynamics cannot be directly extended to the industrial urea melt granulation.

A dynamic circuit simulator has been previously implemented to evaluate the steady-state and dynamic performance of urea granulation plants (Cotabarren et al., 2010). This simulator is constituted by two dynamic models: multi-chamber fluidized-bed granulator (which assumes coating as the main growth mechanism; Bertin et al., 2010a) and fluidized-bed cooler (Cotabarren et al., 2010); and the pseudo-steady-state models earlier described for the crusher and screens (Cotabarren et al., 2008 and Cotabarren et al., 2009). Pseudo-steady-state models were used for the screen and crusher as the dynamics of these processes were much faster than those of the fluidized-bed granulator and cooler, being their effect on the whole circuit dynamics not significant.

One of the applications of simulating the granulation circuit under steady-state conditions is the possibility of carrying out a sensitivity analysis. This allows studying the effect of different operating variables on, among others variables, the PSDs and mass flowrates of the product and recycle streams. Once the critical variables are identified, potential manipulated variables for future optimizations can be recognized, control strategies can be defined, etc.

8.1.5. Selected Steady-State Simulations

In this section, some relevant results of the performed steady-state sensitivity analysis (Cotabarren et al., 2010) are presented to illustrate the effect of different operating variables disturbances on the circuit performance.

This study was carried out by running different steady-state simulations of the entire granulation circuit. The chosen manipulated variables were the urea melt flowrate within each chamber ($\dot{M}_{melt}$), top and bottom decks screen apertures (h_T and h_B) and upper and lower crusher gaps (GAP_U and GAP_L). Each of them was disturbed ± 10 % around its initial steady-state value, which corresponded to a typical condition of an industrial plant operating under nominal capacity. The circuit performance was monitored by tracking different variables, particularly the product and recycle quality, recycle ratio, screen oversize and undersize stream flowrates, granulation chamber heights and temperatures.

The quality of the urea streams was followed by the product and recycle SGN and the product fraction on specification. As aforementioned, values of SGN between 270 and 310 are within the commercial standards for the special case of granular urea. Furthermore, the product size is preferred to be within the range 2 to 4 mm (www.kafcobd.com; www.cfindustries.com; Giovanelli and Schech, 2004).

Figures 26 to 28 present the final steady-state response (as the percentage of variation from the initial steady-state) of relevant granulation circuit variables when different disturbances were applied to the system. The sensitivity analysis reveals that the effect of the variables on the circuit operation verifies the following order of importance: $h_T \approx h_B > GAP_L > \dot{M}_{melt} > GAP_U$. Although the screen top and bottom deck apertures cannot be changed under operation, the study of their influence on the circuit performance can be useful to define appropriate screen meshes or to predict the effect of apertures blinding, being this last operating problem particularly important for the bottom deck that handles a lot of fines.

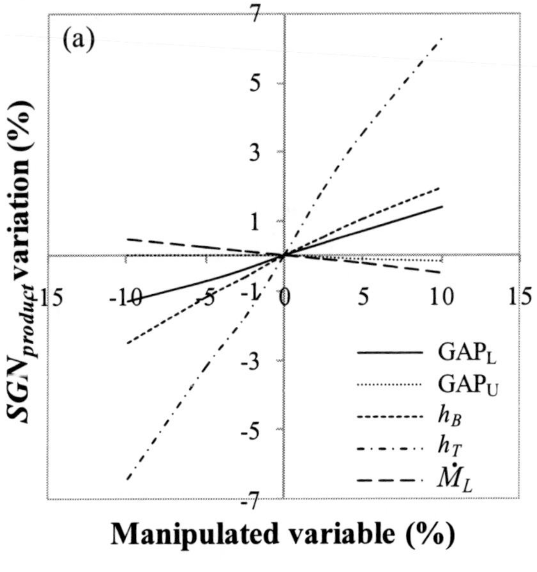

Figure 26.a. Product quality variables: product SGN.

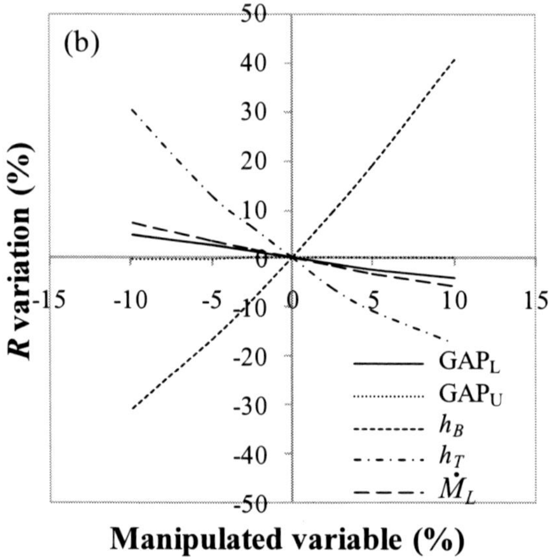

Figure 26.b. Product quality variables: recycle ratio.

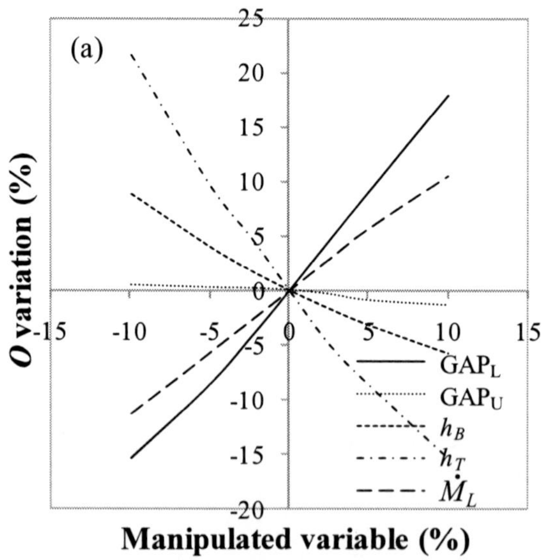

Figure 27.a. Screen variables: oversize mass flowrate.

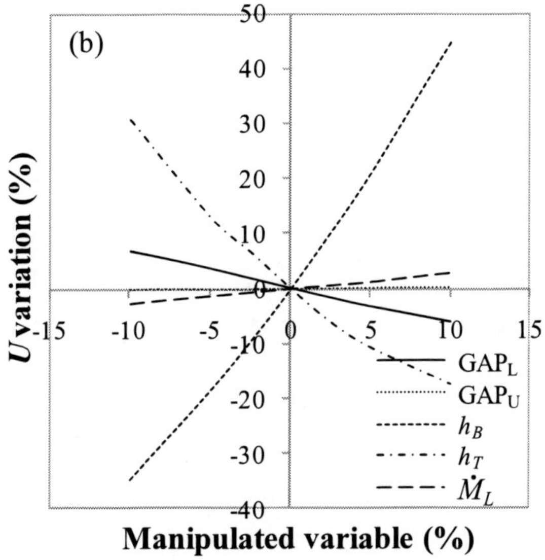

Figure 27.b. Screen variables: undersize mass flowrate.

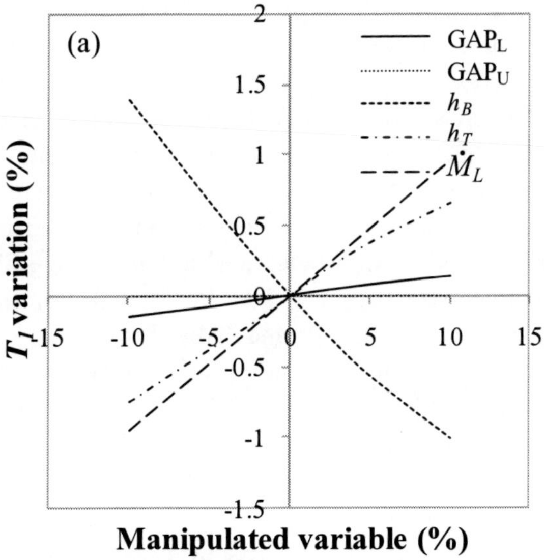

Figure 28.a. Granulator chambers variables: first chamber temperature.

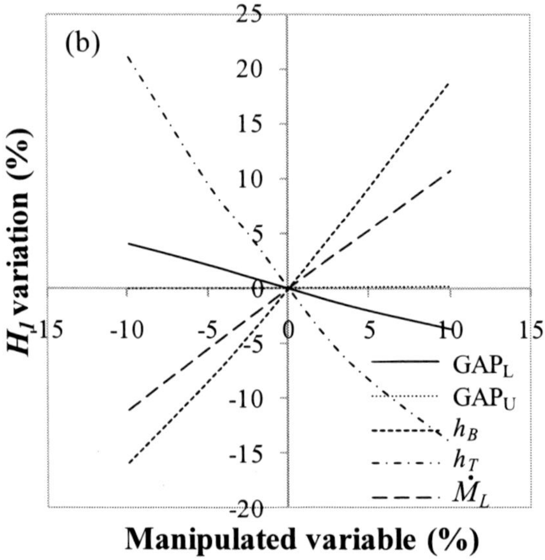

Figure 28.b. Granulator chambers variables: first chamber height.

8.1.6. Open Loop Dynamics: Bottom Deck Screen Disturbances

The simulator of the urea granulation circuit allows studying the dynamics of the system. As an example, the open loop dynamic analysis for a -10% step disturbance in the screen bottom deck (which according to the steady-state analysis is one of the most influencing circuit variables) is shown in Figure 29. Besides, as aforementioned, this disturbance simulates the screen blinding that can occur during real operation. When the bottom deck aperture decreased, immediately smaller and less particles per unit time were allowed to pass through it to the undersize stream (9 and 35% initial drop in the undersize SGN and mass flowrate, Figures 29.a and 29.b). Consequently, finer and more particles per unit time remained in the product stream which at first had a higher mass flowrate (22%) and a lower SGN (3%). Since the undersize mass flowrate is one order of magnitude bigger than the crusher one (data not shown), the recycle behavior almost replicated the trend exhibited by the undersize stream. As soon as the bottom deck aperture decreased, the seeds flow coming into the granulator became lower, being the particles considerably smaller (initially changes of -32% and -9% for the seeds mass flowrate and SGN, respectively) as a consequence of the decrease in the undersize stream SGN and mass flowrate.

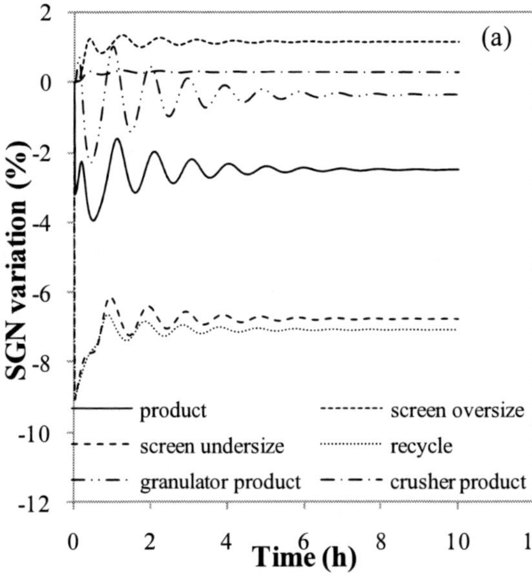

Figure 29.a. Open loop dynamic analysis for a -10% disturbance in the bottom deck aperture: streams SGN.

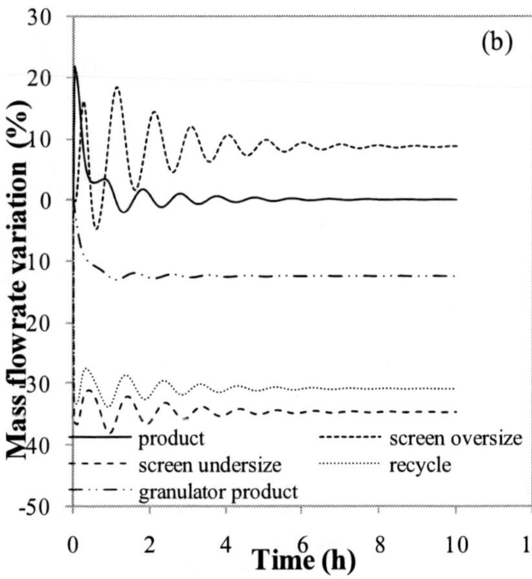

Figure 29.b. Open loop dynamic analysis for a -10% disturbance in the bottom deck aperture: streams mass flowrate.

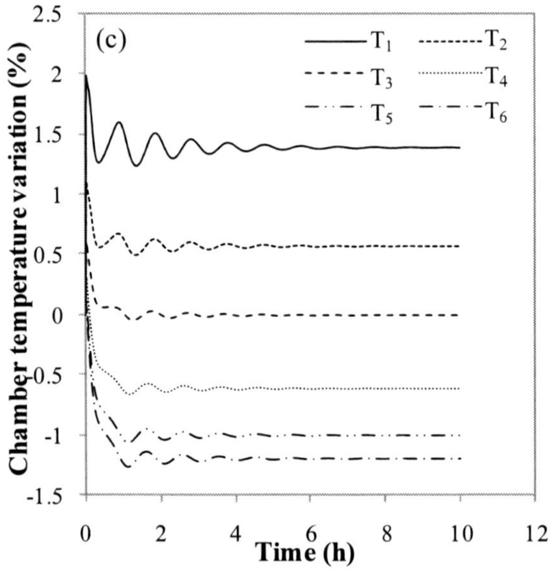

Figure 29.c. Open loop dynamic analysis for a -10% disturbance in the bottom deck aperture: granulator chambers temperatures.

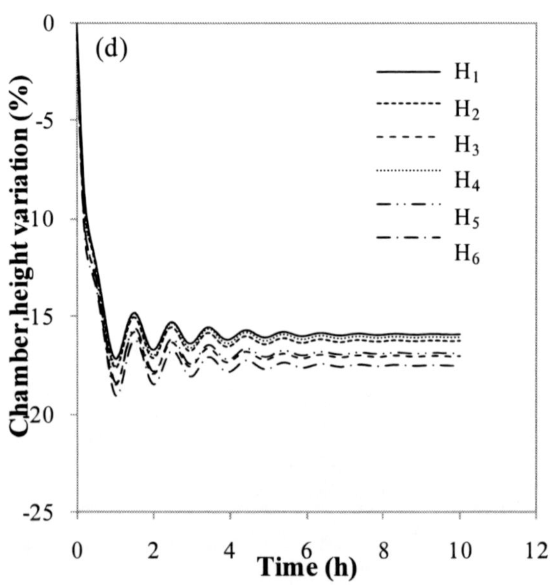

Figure 29.d. Open loop dynamic analysis for a -10% disturbance in the bottom deck aperture: granulator chambers height.

Immediately after the disturbance in the bottom deck aperture, the particles number within the growth chambers started to decline (data not shown). This decrease in the particles number led to less solids superficial area and, therefore, higher particles growth rate. Indeed, the same binder flowrate had to be spread over a lower available surface. Consequently, the granulator product SGN increased but its mass flowrate decreased because less particles per unit time were recycled to the size enlargement unit (Figures 29.a and 29.b). Thus, transitorily, the stream to the top deck had a higher mass median and lower mass flowrate. As a result of the higher coarse fraction, the mass flowrate and SGN of the oversize increased. Instead of the immediate response exhibited by the product, undersize and recycle streams, the oversize and granulator exit streams evolved gradually from the initial steady-state as the disturbance propagated through the whole circuit. Once the screen feed and oversize evolved as described, the screen product and undersize responded to this change with increments in their SGN and diminutions in their flowrates. The cycles took place until all the effects were compensated and the new steady state was reached. Consequently, the behavior of the circuit cannot be predicted without its simulation; in fact both the quality and the mass flow of each solid stream have to be accounted for. Therefore, a tool that represents accordingly the process is essential to study the circuit dynamics.

It is interesting to note that lower mass flowrate into the granulator due to smaller screen bottom deck aperture, diminished the solids mass holdup and, therefore, the fluidized-bed height (Figure 29.d). Together with this effect, the temperature of all the chambers became initially higher, but then only the first and second chamber stabilized in a higher temperature with respect to the initial steady-state (Figure 29.c). During the oscillation cycles, as the bed heights decreased, the temperatures increased and vice versa. As the mass-holdup diminished, the inlet enthalpy provided by a constant melt flowrate per unit of bed mass increased and, thus, the bed temperatures of the growth chambers increased too.

Nevertheless, for the -10% disturbance in the bottom screen deck aperture, the system achieved a final steady state without any control action and the product SGN exhibited relatively low changes (as maximum 4% with respect to the initial steady state). Therefore, the circuit has certain capability of self control. However, the bed mass hold-ups change considerably and, consequently, should be kept under control to ensure the desired accretion growth mechanism (i.e. a decrease in the chambers mass for a constant amount of binder tends to favor agglomeration).

The bottom deck aperture, of course, cannot be modified under normal operation. However, as mentioned, this variable can be subjected to changes over time because of the apertures blinding due the fines handling. Therefore, these results are valuable to predict the plant performance under real operation.

8.1.7. Dynamic Optimization: Optimal Control Study

In this section, an optimal control study performed by using the developed dynamic circuit simulator is presented (Cotabarren et al., 2011). It was already mentioned that there are clear motivations to focus the research on urea granulation circuits to improve the efficiency of large-scale plants and increase their competitiveness. Therefore, the results of the optimal control proposed to maximize plant throughput, by driving the process from an initial stable steady-state (corresponding to a large-scale plant operating at nominal capacity, Cotabarren et al., 2011) to another one by simultaneously satisfying process physical limits as well as manipulated variables constraints during the transient, are briefly described.

During the operation of granulation circuits, a relatively low number of variables can be manipulated either continuously or periodically (among others, urea melt flowrate, gap between rolls in both pairs of the crusher, granulator and cooler fluidization air inlet temperature, and granulator discharge area). All of them can be sequentially considered to maximize the plant capacity and overcome the different bottlenecks that affect the system. As a relevant example, the optimal control with one, two and three sequentially control variables are here analyzed. These variables were: urea melt flowrate and, as suggested by the granulator simulation results presented in Section 6, the granulator discharge area and granulator fluidization air temperature (particularly the air temperature that enters to the hottest chamber, i.e. the second bed).

Regarding the circuit constraints, there are some key limitations. The fluidized-bed chambers heights have a direct influence on the pressure drop, solid mass holdup and degree of droplets net deposition onto the granules surface. These bed heights should be within certain values to guarantee stable operations and to avoid undesired agglomeration and dust formation. In addition, these heights should not exceed the chamber weir one, causing solids overflow or bypass. Besides, the growth bed temperatures have to remain high to avoid melt solidification in the spray nozzles and ensure complete water evaporation. However, too high temperatures (i.e., close to the urea melting point) would lead to partial or total quenching of the bed (Bertin et al., 2010a). For these reasons, the bed heights were set to be within 50 and 90 % of the weirs heights and the temperatures in the growth chambers (first three) were restricted to the range 100 to 120 °C. Once again, the product quality was constrained to guarantee a SGN between 270 and 310. For these optimizations, the following objective function was considered:

$$FO[u(t)] = \max_{u(t),\, t \in [t_0, t_f]} \int_{t_0}^{t_f} P[x(t), u(t)]dt \qquad (6)$$

where FO is the objective function, u represents the control variables, P the urea production to be maximized from the initial to the final operating time (t_0 and t_f, respectively), x the state variables of the process model and t the operation time. The optimal control problem was subjected to the following constraints:

$$270 \leq SGN_{product}(t) \leq 310 \qquad (7)$$

$$0.5H_{weir} \leq H_k(t) \leq 0.9H_{weir} \qquad for\ k = 1\ to\ 6 \qquad (8)$$

$$100°C \leq T_k(t) \leq 120°C \qquad for\ k = 1\ to\ 3 \qquad (9)$$

The optimal control problem also required the definition of the simulation final time as well as the number and duration of the control intervals. According to the disturbances imposed to the circuit shown in the previous section, ten hours was a reasonable time for reaching a new steady state. The minimum duration of the control intervals was set in 10 minutes to ensure a feasible control in the industrial practice. Simulations for different number of control intervals indicated that a number of ten was the minimum one to achieve the optimal objective function value. It has also been proved that time horizons smaller than ten hours but long enough to reach the new steady state, did not modify substantially either the control variables time-trajectories or the optimal objective function values.

Figures 30.a, 30.b and 30.c show the evolution of the control variables when manipulating one (urea melt flowrate), two (urea melt flowrate and granulator discharge area, α) and three variables (urea melt flowrate, granulator discharge area and fluidization air temperature to second chamber,

Ta_2). Figure 30.a demonstrates that with more control variables (i.e. more degrees of freedom) the system evolved to a higher plant capacity through more melt injection (being the maximum throughput improvement of about 30%).

Figure 30.b indicates that the discharge area was set at its maximum allowable value in both, two and three manipulated variables control problems. Finally, the evolution of the fluidization air temperature is presented in Figure 30.c, being this variable set at its minimum allowable value through the entire time horizon.

The evolution of the circuit constraints is shown through Figures 30.d to 30.f. It can be seen that for the optimal control with one manipulated variable (1MV) the second chamber's height was the only active constraint at the beginning of the simulation time (Figure 30.d). This suggested that in order to achieve higher throughputs, it is necessary to use a second control variable capable of regulating the bed heights. In this context, the granulator solids discharge area appeared as an attractive variable to be manipulated. Actually, the granules are discharged from this unit by ducts located at the bottom of the last chamber. At the end of the ducts, a swing type valve is often used to control the solids discharge. The opening angle of this valve allows to modify the discharge area and, thus, to regulate the solids mass flowrate from the granulator.

By adding the granulator discharge area as a second manipulated variable (second optimal control problem, 2MV), the urea production increased but the second chamber´s height was still an active constraint (Figure 30.d) for most of the time horizon. Furthermore, due to the increment in the melt injected to the chambers, the bed temperatures increased considerably and the third chamber temperature reached the upper constraint (Figure 30.e).

Consequently, the fluidization air flowrate became a possible third manipulated variable to solve this bottleneck. For the optimal control problem with three manipulated variables (3MV), the results indicate that only the second chamber height (Figure 30.d) acted as active constraint during most of the simulation time. The product SGN was never an active constraint as shown in Figure 30.f.

The previous optimal control study clearly demonstrated that the manipulation of just the granulator melt flowrate (trivial solution to the plant capacity revamping problem) does not improve in an attractive level the production because of the height physical constraint.

The possibility of incorporating the granulator discharge area as a second control variable highly increases the urea product flowrate. The fluidization air temperature as third control variable allows an even higher throughput, although the increment is not as significant. Furthermore, for this last option, the addition of an efficient cooling air system to reduce the fluidization air inlet temperature is required and a cost analysis should be performed in order to determine if this strategy is economically worthy. Therefore, the manipulation of both urea melt flowrate and granulator discharge area is an attractive strategy.

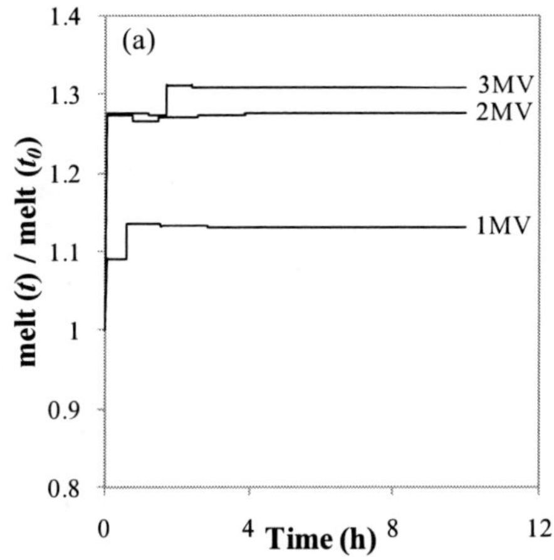

Figure 30.a. Manipulated and process variables time-trajectories: urea melt.

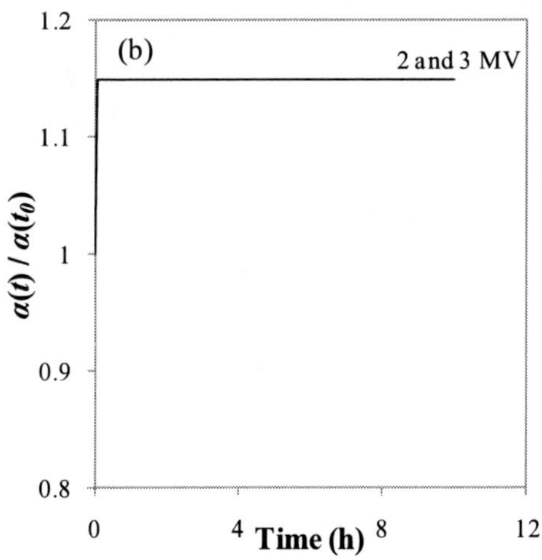

Figure 30.b. Manipulated and process variables time-trajectories: granulator discharge area.

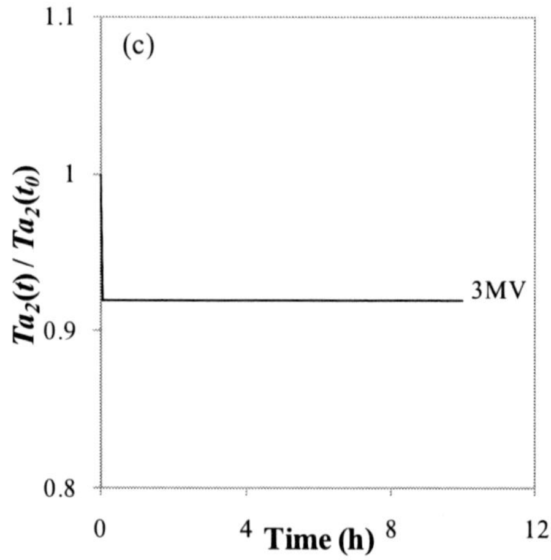

Figure 30.c. Manipulated and process variables time-trajectories: fluidization air temperature.

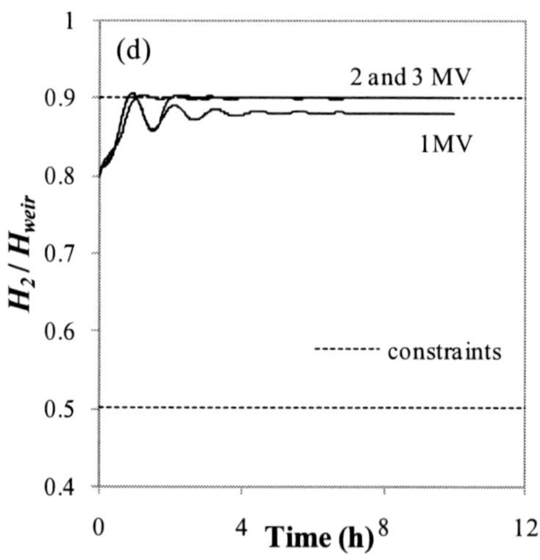

Figure 30.d. Manipulated and process variables time-trajectories: granulator second chamber height.

From the previous steady- and dynamic-state simulations as well as the performed optimal control problems it is clear the importance of having a complete simulator of the granulation circuit in order to predict correctly the transient evolution of the plant, identify circuit bottlenecks and improve the urea granulation circuit performance.

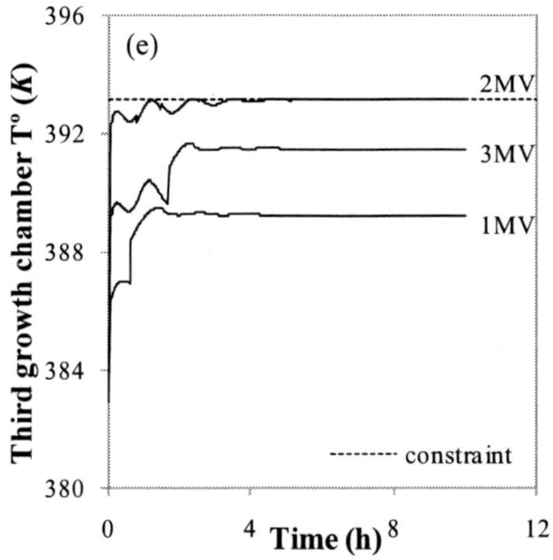

Figure 30.e. Manipulated and process variables time-trajectories: granulator third chamber temperature.

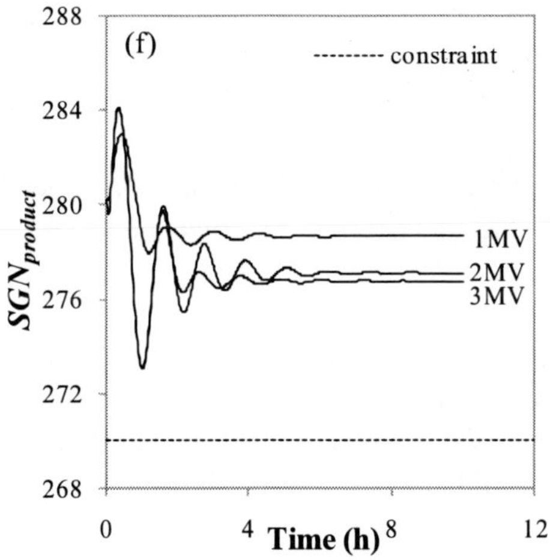

Figure 30.f. Manipulated and process variables time-trajectories: product SGN.

CONCLUSION

Granulation is now superseding prilling as the method of choice for urea solid form production. Today, many urea plants operate with fluidized-bed granulation technologies as finishing process and very large-scale single train plants are running (up to nearly 4000 tpd). For the next years the urea demand is expected to grow and, consequently, new urea plants are planned to come on stream. There are many urea granulation plants spread around the world, which are mainly operated by trial and error. The urea world installed capacity and the forecasted growth for the urea market indicate the need to focus research on urea granulation circuits to improve the efficiency of the plants in order to increase their competitiveness. In that sense, lot of effort has been made to understand the whole urea granulation circuit and to identify the main plant bottlenecks and the most influential operating variables. The theory currently available for urea granulation is basically on a macroscopic level and quite detailed. The behavioral characteristics of urea granulation on a meso-microscopic level are not clearly understood. In this sense, experimental work has been performed in a pilot urea fluidized-bed granulator trying to understand the effect of the operating conditions on the growth mechanisms. General theories from these data have to be built in order to determine which variables need to be adjusted to achieve a desired result, aiming to shift urea granulation from art to science.

REFERENCES

Abberger, T.; Seo, A.; Schæfer, T. The Effect Of Droplet Size and Powder Particle Size on the Mechanisms of Nucleation and Growth in Fluid Bed Melt Agglomeration, *International Journal of Pharmaceutics* 2002, *249*, 185-197.

Adetayo, A.A. *Modelling and Simulation of a Fertilizer Granulation Circuit.* PhD Thesis, University of Queensland, Australia, 1993.

Adetayo, A.; Litster, J.; Cameron, I. T. Steady State Modelling and Simulation of a Fertilizer Granulation Circuit. *Computers and Chemical Engineering* 1995, *19*, 383-393.

Alnajar, A. Framing the Best Practices in Granulation Operation. *Arab Fertilizer* 2010, *58*, 26-29.

Aspen Plus 12.1 Unit Operation Models.; Aspen Technology Inc.: Cambridge, MA, June 2005.

Austin, L.G.; Van Orden, D.; Perez, J. W. A Preliminary Analysis of Smooth Roll Crushers. *International Journal of Mineral Processing* 1980, *6*, 321-336.

Austin, L.G. A Review, Introduction to the Mathematical Description of Grinding as a Rate Process, *Powder Technology* 1971/1972, *5*, 1-17.

Austin, L.G.; Bhatia, V.K. Experimental Methods for Grinding Studies in Laboratory Mills, *Powder Technology* 1971/1972, *5*, 261-266.

Austin, L.G; Cho, H. An Alternative Method for Programming Mill Models, *Powder Technology* 2002, *122*, 96-100.

Baeder, A. Rotoform Urea Pastilles. The Sustainable Premium Urea product. Process Paper 2010; http://www.ureaknowhow.com.

Balliu, N. *An Object Oriented Approach to the Modelling and Dynamics of Granulation Circuits*. PhD Thesis, University of Queensland, Australia, 2005.

Balliu, N.; Cameron, I. T. Performance Assessment and Model Validation for an Industrial Granulation Circuit. *Powder Technology* 2007, *179*, 12-24.

Benzer, H.; Ergün, L.; Öner, M.; Lynch, A.J. Simulation of Open Circuit Clinker Grinding, *Minerals Engineering* 2001, *14*, 701-710.

Bertin, D.E.; Mazza, G.D.; Piña, J.; Bucalá, V. Modeling of an Industrial Fluidized Bed Granulator for Urea Production. *Industrial and Engineering Chemistry Research* 2007, *46*, 7667-7676.

Bertin, D.E.; Piña, J.; Bucalá, V. Dynamics of an Industrial Fluidized-Bed Granulator for Urea Production. *Industrial and Engineering Chemistry Research* 2010a, *49*, 317-326.

Bertin, D.E.; Piña, J.; Bucalá, V. Analysis of a Three-Zone Model for Melt Granulation in a Continuous Fluidized Bed. *World Congress on Particle Technology*, Nuremberg, Germany 2010b.

Bertin, D.E.; Cotabarren, I. M.; Bucalá, V.; Piña, J. Analysis of the Product Granulometry, Temperature and Mass Flow of an Industrial Multichamber Fluidized Bed Urea Granulator, *Powder Technology* 2011, *206*, 122-131.

Boerefijn, R.; Hounslow, M.J. Studies of Fluid Bed Granulation in an Industrial R&D Context, *Chemical Engineering* 2005, *Science 60*, 3879-3890.

Brouwer, M. The Different Forms of Solid Urea and the Background of Caking. Process Paper 2010a; http://www.ureaknowhow.com.

Brouwer, M. World Scale Urea Plants. Process Paper 2010b; http://www.ureaknowhow.com.

Brouwer, M. A Single Solution to the Four Challenges Facing your Urea Prilling Plant. Process Paper 2011; http://www.ureaknowhow.com.

Bucalá V.; Piña J. Teaching Population Balances for Chemical Engineering Students, Application to Granulation Processes, *Chemical Engineering Education* 2007, *41(3)*, 209-217.

Buitnik, F. Recent Highlights of Stamicarbon Urea Technology. *22nd AFA International Fertilizers Technology Conference and Exhibition*, Marrakech, Morocco, June 30 - July 2 2009.

Cameron, I.T.; Wang, F.Y.; Immanuel, C.D.; Stepanek, F. Process Systems Modelling and Applications in Granulation: A Review. *Chemical Engineering Science* 2005, *60(14)*, 3723– 3750.

Campbell, G.; Webb, C. On Predicting Roller Milling Performance Part I: The Breakage Equation. *Powder Technology* 2001, *115*, 234-242.

Capes, C.E. Particle Size Enlargement, In Handbook of Powder Technology, Williams, J.C.; Allen, T. (Eds.), Elsevier Scientific Publishing Company. Amsterdam. 1980.

CF Industries. Deerfield, Illinois, USA; http://www.cfindustries.com/ProductUrea.htm.

Chua, K.W.; Makkawi, Y.T.; Hewakandamby, B.N.; Hounslow, M.J. Time Scale Analysis for Fluidized Bed Melt Granulation-II: Binder Spreading Rate, *Chemical Engineering Science* 2011a, *66(3)*, 327-335.

Chua, K.W.; Makkawi, Y.T.; Hounslow, M.J. Time Scale Analysis for Fluidized Bed Melt Granulation I: Granule-Granule and Granule-Droplet Collision Rates, *Chemical Engineering Science* 2011b, *66*, 318-326.

Cotabarren, I.M.; Schulz, P.G.; Bucalá, V.; Piña, J. Modeling of an Industrial Double-Roll Crusher of a Urea Granulation Circuit. *Powder Technology* 2008, *183(2)*, 224-230.

Cotabarren, I.M.; Rossit, J.; Bucalá, V.; Piña, J. Modeling of an Industrial Vibrating Double-Deck Screen of a Urea Granulation Circuit. *Industrial and Engineering Chemistry Research* 2009, *48*, 3187-3196.

Cotabarren, I.M.; Bertin, D.E.; Romagnoli, J.; Bucalá, V.; Piña, J. Dynamic Simulation and Optimization of a Urea Granulation Circuit. *Industrial and Engineering Chemistry Research* 2010, *49*, 6630-6640.

Cotabarren, I.M.; Bertin, D.E.; Piña, J.; Bucalá, V. Analysis of Optimal Control Problems and Plant Debottlenecking for Urea Granulation Circuits. *Industrial and Engineering Chemistry Research* 2011, *50*, 11996-12010.

Cruz, N.; Briens, C.; Berruti, F. Supersonic Attrition Nozzles in Gas–solid Fluidized Beds, *Chemical Engineering and Processing* 2010, *49*, 225-234.

Daubert, T.E.; Danner, R.P. Thermodynamics Properties of Pure Chemicals, Data Compilation. Taylor and Francis, London, England. 1996.

Dosta, M.; Heinrich, S.; Werther, J. Fluidized Bed Spray Granulation: Analysis of the System Behaviour by Means of Dynamic Flowsheet Simulation. *Powder Technology* 2010, *204*, 71-82.

Drechsler, J.; Peglow, M.; Heinrich, S.; Ihlow, M.; Morl, L. Investigating the Dynamic Behaviour of Fluidized Bed Spray Granulation Processes Applying Numerical Simulation Tools. *Chemical Engineering Science* 2005, *60*, 3817-3833.

Ennis, B.J.; Tardos, G.; Pfeffer, R. A Microlevel-Based Characterization of Granulation Phenomena, *Powder Technology* 1991, *65*, 257-272.

FAO, Food and Agriculture Organization of the United Nations, La Agricultura Mundial en la Perspectiva del Año 2050, Roma, Italia, 2009.

Ferrara, G.; Preti, U.; Schena, G. D. Modelling Screening Operations. *International Journal of Mineral Processing* 1988, *22*, 193-222.

Fertilizer Manual. United Nations Industrial Development Organization (UNIDO) and International Fertilizer Development Center (IFDC). Kluwer Academic Publishers: The Netherlands, 1998.

Finck, A. Fertilizantes y Fertilización, Verlag Chemie GmbH (versión en español), Editorial Reverté S.A., 1988.

Franzrahe, H. Start-up Experience of the MOPCO Urea Granulation Plant. *23rd AFA International Technical Conference and Exhibition*, Tunisia, June 29 – July 1 2010.

Fries, L.; Antonyuk, S.; Heinrich, S.; Palzer, S. DEM–CFD Modeling of a Fluidized Bed Spray Granulator, *Chemical Engineering Science* 2011, *66*, 2340-2355.

Giovanelli, A.; Schech, R. The Stacx Project. *10th Stamicarbon Urea Symposium*, 2004.

Goldschmidt, M.J.V.; Beetstra, R.; Kuipers, J.A.M. Hydrodynamic Modelling of Dense Gas-fluidised Beds: Comparison of the Kinetic Theory of Granular Flow With 3D Hard-sphere Discrete Particle Simulations, *Chemical Engineering Science* 2002, *57*, 2059-2075.

Goldschmidt, M.J.V.; Weijers, G.G.C.; Boerefijn, R.; Kuipers, J.A.M. Discrete Element Modelling of Fluidised Bed Spray Granulation, *Powder Technology* 2003, *138*, 39-45.

Hede, P.D.; Bach, P.; Jensen, A.D. Batch Top-spray Fluid Bed Coating, Scale-up Insight Using Dynamic Heat- and Mass-transfer Modeling, *Chemical Engineering Science* 2009, *64(6)*, 1293-1317.

Heffer, P.; Prud'homme, M. Fertilizer Outlook 2011 - 2015. *79th IFA Annual Conference*, Montreal, Canada, May 2011.

Heinrich, S.; Henneberg, M.; Peglow, M.; Drechsler, J.; Mörl, L. Fluidized Bed Granulation, Analysis of Heat and Mass Transfers and Dynamic Particle Populations, *Brazilian Journal of Chemical Engineering* 2005, *22(2)*, 181-194.

Heinrich, S.; Mörl, L. Fluidized Bed Spray Granulation - A New Model for the Description of Particle Wetting and of Temperature and Concentration Distribution, *Chemical Engineering and Processing* 1999, Process Intensification *38(4-6)*, 635-663.

Heinrich, S.; Peglow, M.; Ihlow, M.; Henneberg, M.; Morl, L. Analysis of the Start-Up Process in Continuous Fluidized Bed Spray Granulation by Population Balance Modelling. *Chemical Engineering Science* 2002, *57*, 4369-4390.

Heinrich, S.; Peglow, M.; Ihlow, M.; Morl, L. Particle Population Modeling in Fluidized Bed-Spray Granulation—Analysis of the Steady State and Unsteady Behavior. *Powder Technology* 2003, *130*, 154-161.

Hemati, M.; Cherif, R.; Saleh, K.; Pont, V. Fluidized Bed Coating and Granulation: Influence of Process-Related Variables and Physicochemical Properties on the Growth Kinetics, *Powder Technology* 2003, *130*, 18-34.

Iveson, S.M.; Wauters, P.A.L.; Forrest, S.; Litster, J.D.; Meesters, G.M.H.; Scarlett, B. Growth Regime Map for Liquid-bound Granules, Further Development and Experimental Validation, *Powder Technology* 2001, *117(1-2)*, 83-97.

Karnaphuli Fertilizer Company Limited. Rangadia, Anowara Chittagong, Bangladesh; http://www.kafcobd.com/html/products.html.

Karra, V.K. Development of a Model for Predicting the Screening Performance of a Vibrating Screen. *CIM Bulletin* 1979, 167-171.

Kars, F. Prilling or Granulation of Urea. *IFA Technical Conference*, Paris, France, 5-8 November 1984

Kayaert, A.F. *Proceedings of the 30th Annual Meeting Fertilizer Industry Round Table*. Atlanta, Georgia, 1980.

Kayaert, A.F.; Antonus, R.A.C. Process for the Production of Urea Granules. CA Patent 2141874. 1994.

Kayaert, A.F.; Antonus, R.A.C. Process for the Production of Urea Granules. US Patent 5653781. 1997.

Kis, P.B.; Mihálykó, C.; Lakatos, B.G. Discrete Model for Analysis and Design of Grinding Mill-Classifier Systems, *Chemical Engineering and Processing* 2006, *45*, 340-349.

Landis, N. Process for Prilling Urea. U.S. Patent 4190622. 1980.

Li, J.; Webb, C.; Pandiella, S.; Campbell, G. Discrete Particle Motion on Sieves-A Numerical Study Using the DEM Simulation. *Powder Technology* 2003, *133*, 190-202.

Litster, J.; Ennis, B.; Liu, L. The Science and Engineering of Granulation Processes. Particle Technology Series, Kluwer Academic Publishers, Dordrecht. 2004.

Mavrovic, I. Process for Prilling Urea. U.S. Patent 3933956. 1976.

Meessen J.H.; Petersen H. Urea. In: Ullmann's Encyclopedia of Industrial Chemistry, ed. VCH, Weinheim, Germany, 1996.

Meessen, J.; van Baal, H. Stamicarbon's Granulation Technology: Diamonds, Pearls and Stamicarbon Granules. *IFA Technical Committee*, Hilton Abu Dhabi. 30 September – 2 October 2003.

Mikkelsen, R.L. Biuret in Urea Fertilizers. *Better Crops with Plant Food* 2007, *91(3),* 6-7.

ModsimSimulator; Mineral Technologies Inc.: Salt Lake City, UT; http://www.mineraltech.com/MODSIM. 2008.

Moly-CopTools; Moly-Cop Chile S.A.: Santiago, Chile; http://www.molycop.cl. 2008.

Nagaiah, C.; Warnecke, G.; Heinrich, S.; Peglow, M. Three-dimensional Numerical Study of Heat and Mass Transfer in Fluidized Beds with Spray Nozzle, *Computers and Chemical Engineering* 2008, *32(12)*, 2877-2890.

Nakamura, S. The TOYO Urea Granulation Technology - The Challenges and Achievements in Producing Urea Granules. *20th AFA International Annual Technical Conference*, Tunisia, 19-21 June, 2007.

Niehues, P.; Antonus, R. Yara's Well Established Fluid Bed Granulation Now at Uhde Fertilizer Technology. *Uhde Fertiliser Symposium 2006*, Dortmund, Germany, 17-19 May 2006.

Nijsten, P.J.B.; Starmans, P.J.M. Process for Producing Granules, U.S. Patent 5779945. 1998.

Niks, A.; Van Hijfte, W.H.P.; Goethals, R.A.J. Process for Urea Granulation, US Patent 5653781. 1997.

Niks, A.; Van Hijfte, W.H.P.; Goethals, R.A.J. Process for Urea Granulation. U.S. Patent 4219589. 1980.

Pietsch, W. Size Enlargement by Agglomeration, John Wiley and Sons Ltd, Chichester, England. 1991.

Pont, V.; Saleh, K.; Steinmetz, D.; Hemati, M. Influence of the Physicochemical Properties on the Growth of Solid Particles by Granulation in Fluidized Bed, *Powder Technology* 2001, *120*, 97-104.

Poon, J.M.-H.; Ramachandran, R.; Sanders, C.F.W.; Glaser, T.; Immanuel, C.D.; Doyle III, F.J.; Litster, J.D.; Stepanek, F.; Wang, F.-Y.; Cameron, I.T. Experimental Validation Studies on a Multi-dimensional and Multi-scale Population Balance Model of Batch Granulation, *Chemical Engineering Science* 2009, *64*, 775-786.

Potthoff, M. Mega Urea Granulation Plants up and Running - Potentials for Future Developments. *20th AFA International Technical Fertilizer Conference*, 2007.

Prescott, J.K.; Barnum, R.A. On Powder Flowability, *Pharmaceutical Technology* 2000, *10*, 60-84.

Radichkov, R.; Muller, T.; Kienle, A.; Heinrich, S.; Peglow, M.; Morl, L. A Numerical Bifurcation Analysis of Continuous Fluidized Bed Spray Granulation with External Product Classification. *Chemical Engineering and Processing* 2006, *45*, 826-837.

Rajniak, P.; Stepanek, F.; Dhanasekharan, K.; Fan, R.; Mancinelli, C.; Chern, R.T. A Combined Experimental and Computational Study of Wet Granulation in a Wurster Fluid Bed Granulator, *Powder Technology* 2009, *189(2)*, 190-201.

Ramkrishna, D. Population Balances, Academic Press, London, England. 2000.

Randolph, A.D.; Larson, M.A. *Theory of Particulate Processes*, Academic Press, New York, USA, 1971.

Rhodes, M.J. *Introduction to Particle Technology*; Wiley and Sons: West Sussex, England, 1998.

Ronsse, F.; Pieters, J.G.; Dewettinck, K. Combined Population Balance and Thermodynamic Modelling of the Top-spray Fluidised Bed Coating Process. Part I-Model Development and Validation, *Journal of Food Engineering* 2007, *78(1)*, 296-307.

Ronsse, F.; Pieters, J.G.; Dewettinck, K. Modelling Side-effect Spray Drying in Top-spray Fuidised Bed Coating Processes, *Journal of Food Engineering* 2008, *86*, 529-541.

Roos, W. Granulation: Diamonds, Pearls and Stamicarbon Granules. *11th Stamicarbon Urea Symposium*, Noordwijk, 19-22 May 2008. Paper 17w.

Roy, P.; Khanna, R.; Subbarao, D. Granulation Time in Fluidized Bed Granulators, *Powder Technology* 2010, *199*, 95-99.

Sakata, E.; Genshi, N.; Haruyuki, M. Process for Preparing Granular Urea Product, US Patent 0013092, 2007.

Sakata, E.; Yanagawa, T. Latest Urea Technology for Improving Performance and Product Quality. Technology and Knowledge Research Center, Toyo Engineering Corporation; http://www.toyo-eng.co.jp.

Saleh, K.; Guigon, P. Coating and Encapsulation Processes in Powder Technology. In Handbook of Powder Technology; Salman, A. D.; Hounslow, M. J.; Seville, J. P. K. (Eds.); Elsevier: Amsterdam, 2007.

Seo, A.; Holm, P.; Schæfer, T., Effects of Droplet Size and Type of Binder on the Agglomerate Growth Mechanisms by Melt Agglomeration in a Fluidised Bed, *European Journal of Pharmaceutical Sciences* 2002, *16*, 95-105.

Sharma R.K.; Patel L.R. Study and Evaluation of Hygroscopic Behavior of Phosphatic Fertilizers, ANP y CAN by Using Different Anticaking Agents at 80/90 % RH. A case Study at G.N.F.C. *IFA Technical Conference*, New Orleans, Lousiana, USA, 1 - 4 October 2000.

Smith, P.G.; Nienow, A.W. Particle Growth Mechanisms in Fluidised Bed Granulation-I. The Effect of Process Variables, *Chemical Engineering Science* 1983, *38(8)*, 1223-1231.

Snamprogetti - Saipem, The Urea Technology, 2007; http://www.saipem.eni.it.

Soldinger, M. Interrelation of Stratification and Passage in the Screening Process. *Minerals Engineering* 1999, *12*, 497-516.

Stamicarbon Brochure. Growth by Urea Granules - Transforming the Urea Process with Continuous Innovation; http://www.stamicarbon.com.

Stamicarbon's Contribution to Controlling Food Prices: Low Cost Urea Production, *IPF Conference*, Tehran, Iran, May 2011; http://www.stamicarbon.com.

Subasinghe, G. K. N. S.; Schaap, W.; Kelly, E. G. Modelling the Screening Process: A Probabilistic Approach. *Powder Technology* 1989, *59*, 37-44.

Subasinghe, G. K. N. S.; Schaap, W.; Kelly, E. G. Modelling the Screening Process as a Conjugate Rate Process. *International Journal of Mineral Processing* 1990, *28*, 289-300.

Tan H.S.; Salman A.D.; Hounslow M.J., Kinetics of Fluidised Bed Melt Granulation I: The Effect of Process Variables, *Chemical Engineering Science* 2006, *61*, 1585-1601.

Tardos, G.I.; Irfan Khan, M.; Mort, P.R. Critical Parameters and Limiting Conditions in Binder Granulation of Fine Powders, *Powder Technology* 1997, *94*, 245-258.

Toyo Engineering Corporation; www.toyo-eng.co.jp.

Turton, R.; Tardos, G.I.; Ennis, B.J. Fluidized Bed and Granulation, In Fluidization, Solids Handling and Processing, Industrial Applications, Yang, W. (Ed.), Noyes Publications, New Jersey, USA. 1998.

UFT Fluid-bed Granulation – Superior Technology 2011 Brochure, FL240e/2000 2/2011 Uhde/TK Printmedia/ Germany; http://www.uhde-fertilizer-technology.com.

Uhde Fertilizer Technology; http://www.uhde-fertilizer-technology.com.

Uhde Technology Profile, Uhde GmbH, 2002; http://www.uhde.eu/en.html.

van Sint Annaland, M.; Deen, N.G.; Kuipers, J.A.M. Multi-level Computational Fluid Dynamics Models for the Description of Particle Mixing and Granulation in Fluidized Beds. In Handbook of Powder Technology; Salman, A. D.; Hounslow, M. J.; Seville, J. P. K. (Eds.); Elsevier: Amsterdam, The Netherlands, 2007.

Walker, G.M.; Andrews, G.; Jones, D. Effect of Process Parameters on the Melt Granulation Pharmaceutical Powders, *Powder Technology* 2006, *165*, 161-166.

Walker, G.M.; Holland, C.R.; Ahmad, M.M.N.; Craig, D.Q. Influence of Process Parameters on Fluidised Hot-Melt Granulation and Tablet Pressing of Pharmaceutical Powders, *Chemical Engineering Science* 2005, *60*, 3867-3877.

Wang, V.; Brouwer, M. Worldwide Urea Plants Overview. Process Paper 2011; http://www.ureaknowhow.com.

Werther, J.; Reppenhagen, J. Attrition, In Handbook of Fluidization and Fluid-particle Systems, Yang, W. (Ed.), Marcel Dekker, Inc.: New York, USA. 2003.

Whiten, W.J. *Simulation and Model Building for Mineral Processing*. PhD Thesis, University of Queensland, Australia, 1972.

Wildeboer, W.J. *Steady State and Dynamic Simulations of a Closed Loop Granulation Circuit*. PhD Thesis, Delft University of Technology, 1998.

Wills, B.; Napier-Munn, T. J. *Will's Mineral Processing Technology, Seventh Edition: an Introduction to the Practical Aspects of Ore Treatment and Mineral Recovery*; Butterworth-Heinemann: London, England, 2006.

Xu, B.H.; Yu, A.B. Numerical Simulation of the Gas–solid Flow in a Fluidized Bed by Combining Discrete Particle Method with Computational Fluid Dynamics, *Chemical Engineering Science* 1997, *52*, 2785-2809.

Zhai, H.; Li, S.; Andrews, G.; Jones, D.; Bell, S.; Walker, G. Nucleation and Growth in Fluidised Hot Melt Granulation, *Powder Technology* 2009, *189*, 230-237.

Zhang, J.; Litster, J.; Wang, F. Y.; Cameron, I. T. Evaluation of Control Strategies for Fertiliser Granulation Circuits Using Dynamic Simulation. *Powder Technology* 2000, *108*, 122-129.

Zhu, H.P.; Zhou, Z.Y.; Yang, R.Y.; Yu, A.B. Discrete Particle Simulation of Particulate Systems: A Review of Major Applications and Findings, *Chemical Engineering Science* 2008, *63*, 5728-5770.

In: Nitrogen Fertilizer
Editor: Karlene Winslow

ISBN: 978-1-63321-585-6

Chapter 3

INFLUENCE OF APPLICATION SYMBIOTIC (*B. JAPONICUM*), ASSOCIATIVE (*A. BRASILENSE*) AND NON SYMBIOTIC NITROGEN FIXING BACTERIA (*A. CHROOCOCCUM*) ON THE YIELD AND THE QUALITY OF SOYBEAN GRAIN

Suzana Kristek**, *Andrija Kristek, Željko Kraljičak, Jurica Jović and Irena Antinac

University of J. J. Strossmayer, Faculty of Agronomy, Osijek, Republic of Croatia

ABSTRACT

The excessive use of mineral fertilizers increases the production costs of agricultural crops, while also causing the acidification of soil and deterioration of chemical and microbiological properties. In the case of mineral nitrogenous fertilizers the soil is very unstable and is easily washed off so that eutrophication of groundwater occurs. The aim of this

* Corresponding author: Suzana Kristek. University of J. J. Strossmayer, Faculty of Agronomy, Kralja P. Svačića 1d, 31000 Osijek, Republic of Croatia. E-mail: skristek@pfos.hr.

study was to investigate the possibility of reduction, as well as the complete exclusion, of mineral nitrogen fertilizer in soybean production through the application of: symbiotic bacteria *Bradyrhizobium japonicum*, associative bacteria *Azospirillum brasilense* and non symbiotic bacteria *Azotobacter chroococcum*. Reduction of nitrogen fertilizer through nitrogen-fixing bacteria substitution doesn't only have an economic importance in agricultural production, but is also of great importance to the ecology of the environment.

During two-year investigations on two different soil types (Humogley, Eutric Cambisols) in 4 repetitions and 9 different variants, as follows: 1. control (untreated seed + mineral nitrogen fertilization based on soil analysis); 2. seed treated with symbiotic bacteria *B. japonicum* + mineral nitrogen fertilization based on soil analysis; 3. seed treated with symbiotic bacteria *B. japonicum* + reduced mineral nitrogen fertilization by 30%; 4. seed treated with symbiotic bacteria *B. japonicum* + reduced mineral nitrogen fertilization by 50%; 5. seed treated with symbiotic bacteria *B. japonicum* + reduced mineral nitrogen fertilization by 70%; 6. seed treated with symbiotic bacteria *B. japonicum,* associative bacteria *A. brasilense* and non symbiotic bacteria *A. chroococcum* + mineral nitrogen fertilization based on soil analysis; 7. seed treated with symbiotic bacteria *B. japonicum,* associative bacteria *A. brasilense* and non symbiotic bacteria *A. chroococcum* + reduced mineral nitrogen fertilization by 30%; 8. seed treated with symbiotic bacteria *B. japonicum,* associative bacteria *A. brasilense* and non symbiotic bacteria *A. chroococcum* + reduced mineral nitrogen fertilization by 50%; 9. seed treated with symbiotic bacteria *B. japonicum,* associative bacteria *A. brasilense* and non symbiotic bacteria *A. chroococcum* + reduced mineral nitrogen fertilization by 70%.

All seed inoculated variants accomplished significantly higher results in comparison with control variants, on both soil types and in both years of investigation.

In both years of investigation on Humogley, a soil type with greater pedological, physical, chemical and microbiological characteristics, the best results of investigated parameters were obtained in the seed variant inoculated with all three different species of benefit bacteria (*B. japonicum, A. brasilense, A. chroococcum*) and with reduced mineral nitrogen fertilization by 70%.

On Eutric Cambisols, in the first year of investigation, best results of investigated parameters were obtained in the seed variant inoculated with all three different species of benefit bacteria and with reduced mineral nitrogen fertilization by 50%. In the second year of investigation, which was marked by a lack of rainfall during July and September, at individual investigated parameters there were no statistically significant differences between listed fertilization and fertilization reduced by 30%.

INTRODUCTION

The current agricultural production endeavors include a reduction of costs for production, utilizing natural energy resources, and avoiding environmental contamination. Agricultural production is almost impossible without the introduction of mineral fertilizers, which can adversely affect the environment. Renewable resources provides a viable option to the adverse effects fertilizer production has on the environment. In this regard, nitrogen extraction is a focal point to change current methods. Nitrogen most visible influences the production of organic plant matter and yield, but it is also the most expensive element. On the other hand, microorganim in the ground carry one of the most important process: biological nitrogen fixation.

Biological nitrogen fixation is one of the methods where nitrogen is converted into forms available to plants. This process is clean, fix $190x10^6$ t nitrogen ha^{-1} (Babeva and Zenova, 1989) and is very important for biosphere conservation as a whole (Bray, 1983). Diazotrophs are microorganisms with the ability to fix atmospheric nitrogen and convert it in forms available to plants.

Seed bacterization with symbiotic diazotrophs has become a common measure in the production of legumes. This measure reduces the utilization of nitrogen mineral fertilizers. Additionally, the measure preserves and increases both nitrogen amounts in the soil and the yield quality and quantity. The application of associative diazotrophs is also of great importance in the production of nonlegume plants, as these types of diazotrophs affects the reduction of mineral nitrogen. It can then produce physiologically active substances and increase plant resistance to phytopathogenic fungi in addition to its role in the reduction of pollutants in the soil. Their incorporation into the soil increases the elements of biogeny and soil productivity (Burns, 1995; Hardy et al., 1995; Cvijanović et al., 2007).

The competition for nodulation is a complex phenomenon depending on soil para-meters and genetic traits of both the *Rhizobium* symbiont and the host (Triplett and Sadowsky, 1992). Milić et al. (2002) state that soybean inocula-tion is known to be dependent upon the quality of rhizobia based on their efficiency in N_2 fixation, competition ability, and compatibility to soybean genotypes. Furthermore, several authors (Sadowski et al., 1987; Mrkovački et al., 1992; Hungria and Bihrer, 2000) have studied the differences between *Bradyrhizobium japonicum* strains regarding their effectiveness with different soybean genotypes and their results have shown the differences of nodulation, yield, and total nitrogen accumulated in grains.

Kristek (2001) alleges that the efficiency of symbiotic fixation depends on the pH of the soil, the content of molybdenum in the soil and compatibility between soybean varieties and strains of *Bradyrhizobium japonicum* in the results of her research.

Bacteria of the genera *Azotobacter* and *Azospirillum* are free-living, N-fixing organisms that live in close association with plants in the rhizosphere. Under appropriate conditions, these bacteria can enhance plant development and promote the yield of several agriculturally important crops in different soils and climatic regions (Jagnow, 1987; Becking, 1992; Okon and Labandera-González, 1994). These benefits of *Azotobacter* and *Azospirillum* on plants are attributed to an improvement in root development, an increase in the rate of water and mineral uptake by roots, displacement of fungi and plant pathogenic bacteria, and biological N_2 fixation to a lesser extent (Brown, 1974; Okon and Itzigsohn, 1995).

Positive effects of combined inoculation with *Rhizobium* plus *Azotobacter* or *Azospirillum* strains have been reported for different forage and grain legumes, and were related to the favourable influence of the free-living diazotrophic bacteria on nodule weight and number, N_2 fixation, plant dry-matter accumulation, and N_2 content (Burns et al., 1981; Yahalom et al., 1987; Rodelas et al., 1996). It has been reported that *Azospirillum* strains enhance the uptake of P and/or K by corn, wheat, sorghum, and rice that exert changes on the balance of macro - and micronutrients in wheat and soybean. However, such effects on mineral nutrition can vary significantly among different plant-strain combinations (Lin et al., 1983; Pacovsky et al., 1985; Murty and Ladha, 1988; Bashan et al., 1990; Rodelas et al., 1999).

The increase in mineral uptake (N, P, and K) and microelements may be due to the increase in root length rate, which improves after dual inoculation (Dobbelaere and Okon, 2007). This leads to improved mineral nutrition of the plant concerning molybdenum. It also improves phosphorus and iron uptake to *Rhizobium,* which are beneficial for nodule formation and nitrogen fixation activities (Burdman et al., 1998).

The essential effect of co-inoculation on increasing susceptibility to *Rhizobium* infection may be that *Azospirillum* stimulates the formation of a large number of epidermal cells that differentiate into infectable root hairs (Yahalom et al., 1987).

In a similar manner, Okon (1985) indicated that the improvement of root proliferation, water status and mineral uptake of plant was caused by *Azospirillum*. The increase in plant growth and nitrogen uptake was observed due to the mixed inoculation of *Rhizobium* and *Azospirillum.*

Such an increase in nodulation and plant growth in soybean crop was attributed to the production of growth promoting substances by *Azospirillum* (Tien et al., 1979). The beneficial effect of *Azospirillum* added to a combination with *Rhizobium* was *Azospirillum's* ability to convert tryptophan into Indole Acetic Acid, a substance that promotes the infection process of *Rhizobium* in root hairs (Marre, 1976). On the other hand, cells of *Azospirillum* and *Azotobacter* may have some compounds that could induce new root hair formation and subsequent infection (Okon and Kapulnik, 1986; Anandham et al., 2007; Abdul Jabar and Mohd Saud, 2012).

In this study, selection of PGPR strains that are effective in Croatia as well as Central and Eastern Europe soils were utilized to co-inoculate the soybean with bradyrhizobia. Furthermore, we determined the possibility of reducing nitrogen fertilizer on two soil types that have different chemical and microbiological properties.

Materials and Methods

The experiment was set up on two types of the soil: Humogley and Eutric Cambisols (Table 1) by randomized block design in 4 repetitions and 9 different variants, as follows:

A) Treating seeds with bacteria:
   1 control (only with nitrogen fertilization based on soil analysis)
   2 seed treated with *B. japonicum*
   3 seed treated with *B. japonicum, A. brasilense* and *A. chroococcum*

B) Nitrogen fertilization:
   1 mineral nitrogen fertilization based on soil analysis
   2 reduced mineral nitrogen fertilization by 30%
   3 reduced mineral nitrogen fertilization by 50%
   4 reduced mineral nitrogen fertilization by 70%

Investigations were conducted during 2011 and 2012, on experimental plots of 100 m^2. Soybean cultivar used in the experiment was Ika (Agricultural Institute in Osijek, Croatia). Plant density was approximately 600000 plants per hectare with spacing between the rows of 50 cm. In both years of the investigation winter wheat was used as the preculture.

Basic fertilization was conducted in late September, and sowing fertilization was conducted in mid April (Table 2).

Table 1. Soil characteristics

Investigated properties in a field	Type of soil	
Layer (0 – 0.3 m)	Humogley	Eutric Cambisols
pH (H_2O)	7.63	7.13
pH (KCl)	6.80	6.25
Humus (%)	3.70	2.23
P (mg/ 100 g soil)	24.90	21.80
K (mg/ 100 g soil)	25.24	23.90

Table 2. The fertilization according variants of investigation

Variants of fertilization	Type of soil											
	Humogley						Eutric Cambisols					
	Basic fertilization (kg ha^{-1})			Initial fertilization (kg ha^{-1})			Basic fertilization (kg ha^{-1})			Initial fertilization (kg ha^{-1})		
	N_2	P_2O_5	K_2O	N_2	P_2O_5	K_2O	N_2	P_2O_5	K_2O	N_2	P_2O_5	K_2O
I	25	60	95	45	-	-	40	90	112	60	-	-
II	19.6	60	95	29.4	-	-	28	90	112	42	-	-
III	14	60	95	21	-	-	20	90	112	30	-	-
IV	8.4	60	95	12.6	-	-	12	90	112	18	-	-

I - fertilization based on soil analysis; II - reduced mineral nitrogen fertilization by 30%; III - reduced mineral nitrogen fertilization by 50%; IV - reduced mineral nitrogen fertilization by 70%.

The basic nitrogen fertilizer was applied in the form of urea (46% N_2), phosphorus in the form of triple superphosphate (45% P_2O_5), and potassium in the form of potassium chloride (60% K_2). Before addition of sowing fertilization, nitrogen was applied in the form of calcium ammonium nitrate (27% N_2).

Seed inoculation was conducted before sowing using the following bacterial strains: *Bradyrhizobium japonicum* (DSM No. 30131, type strain), *Azospirillum brasilense* (DSM No. 1690, type strain) and *Azotobacter chroococcum* (DSM No. 2286, type strain). *B. japonicum* were grown in Rhizobium medium (Yeast extract 1.00 g, Mannitol 10.00 g, Agar 15.00 g, Soil extract 200.00 ml, Distilled water 800.00 ml) at 26 °C. *A. brasilense* were grown in Azospirillum medium (Yeast extract 0.05 g, K_2HPO_4 0.25 g, $FeSO_4 \times 7 H_2O$ 0.01 g, $Na_2MoO_4 \times 2 H_2O$ 1.00 mg, $MnSO_4 \times H_2O$ 2.00 mg, $MgSO_4 \times 7 H_2O$ 0.20 g, NaCl 0.10 g, $CaCl_2 \times 2 H_2O$ 0.02 g, $(NH_4)_2SO_4$ 1.00 g, Biotin 0.10 mg, Bromothymol blue 25.00 mg, Distilled water 950.00 ml) at 30 °C.

Bromothymol blue was diluted in KOH before adding it to the medium. pH of the medium was adjusted to 7.1 and then autoclaved for 15 min at 121 °C. After sterilization we added 25 ml of both filter-sterilized 20% glucose and 20% Na-malate. *A. chroococcum* were grown in Azotobacter medium (Glucose 5.00 g, Mannitol 5.00 g, $CaCl_2$ x 2 H_2O 0.10 g, $MgSO_4$ x 7 H_2O 0.10 g, Na_2MoO_4 x 2 H_2O 5.00 mg, K_2HPO_4 0.90 g, KH_2PO_4 0.10 g, $FeSO_4$ x 7 H_2O 0.01 g, $CaCO_3$ 5.00 g, Agar 15.00 g, Distilled water 950.00 ml) at 30 °C.

Bacterial strains were applied to sterilized peat. Biopreparation humidity levels were at 55% and consisted of *B. japonicum* at 9 x 10^{10} CFU per g of peat, or all three bacterial species (*B. japonicum*, *A. brasilense* and *A. chroococcum*) each at 3 x 10^{10} CFU per g of peat. We confirmed the positive synergies between bacterial strains in laboratory conditions. Also, during the previous two years of research we demonstrated the compatibility between the soybean varieties and *B. japonicum* strains used.

Second year of investigation was recorded with bad weather condition for soybean production, primarily because of the lack of rainfall (Table 3).

When about 80% of the soybean plants were in the flowering stage the average number of nodules and dry weight of nodules were determined. After soybean harvest was determined the soybean grain yield, the average number of pods per plant and the average number of grains per pod.

Table 3. The monthly precipitation and mean monthly air temperature in vegetation of soybean during 2011 and 2012 years as well as perennial average for Osijek

Month	The monthly precipitation (mm)			Mean monthly air temperature (°C)		
	2011	2012	*Average 1961 – 1990*	2011	2012	*Average 1961 – 1990*
V	80.1	54.2	*58.5*	16.7	16.9	*16.5*
VI	78.2	60.8	*88.0*	20.7	22.4	*19.4*
VII	73.9	41.9	*64.8*	22.2	24.7	*21.1*
VIII	55.3	4.1	*58.5*	23.1	24.1	*20.3*
IX	46.1	21.8	*44.8*	17.3	20.9	*16.6*
Total (in vegetation)	333.6	182.8	*314.6*	–	–	–
Average (in vegetation)	–	–	–	20.0	21.08	*18.8*

Grain protein content was determined by the Kjeldahl method. The sample size was 10 plants. The obtained results were processed using statistical methods (analysis of variance, F-test, LSD test).

RESULTS AND DISCUSSION

Based on our results of the investigated parameters (soybean grain yield, average number of pods per plant, average number of grains per pod, average number of nodules, average dry weight of nodules, average grain protein content) we show a significant effect of seed inoculation to control ratio in both years of investigation and on both soil types.

Soybean Grain Yield

During the first year of investigation on soil type Humogley, the highest average yield of soybean (Table 4) was achieved in variant 9 (seed treated with symbiotic bacteria *B. japonicum,* associative bacteria *A. brasilense* and non symbiotic bacteria *A. chroococcum* + reduced mineral nitrogen fertilization by 70%), but there were no statistically significant differences ($p>0.01$) between the listed variant and variant 8 (seed treated with symbiotic bacteria *B. japonicum,* associative bacteria *A. brasilense* and non symbiotic bacteria *A. chroococcum* + reduced mineral nitrogen fertilization by 50%). All other variants achieved a statistically significant ($p<0.01$) lower average grain yield. When seeds were treated only with the bacterium *B. japonicum*, the highest yield achieved was with variant 4 (reduced mineral nitrogen fertilization by 50%). However, there were no statistically significant differences between the variant 3 (reduced mineral nitrogen fertilization by 30%) and variant 5 (reduced mineral nitrogen fertilization by 70%).

Unfavorable drought conditions in the second year of research resulted in poor growth. The highest average soybean yield was again observed in variant 9. All other variants of research achieved statistically significant ($p<0.01$) lower average yield of soybean.

The highest average yield in Eutric Cambisols soil for both years resulted in variant 7 (seed treated with symbiotic bacteria *B. japonicum,* associative bacteria *A. brasilense* and non symbiotic bacteria *A. chroococcum* + reduced mineral nitrogen fertilization by 30%) and 8 (seed treated with symbiotic bacteria *B. japonicum,* associative bacteria *A. brasilense* and non symbiotic bacteria *A. chroococcum* + reduced mineral nitrogen fertilization by 50%). All other variants of research achieved a statistically high significant ($p<0.01$) lower average grain yield.

It is noteworthy that in both years of investigation and on both soil types there was no statistically significant difference between variant 1 – control (untreated seed + mineral nitrogen fertilization based on soil analysis) and variant 2 (seed treated with symbiotic bacteria *B. japonicum* + mineral nitrogen fertilization based on soil analysis). This is consistent with the results of many authors who suggest that lower levels of nitrogen fertilization and content of organic matter in the soil stimulates nodulation and biological nitrogen fixation. This directly increases the value of all elements of yield, while high levels of nitrogen inhibit all elements of yield (Essa, 1985; Ergorov, 1985; Waterer et al., 1992; Thies et al., 1995; Kristek, 2001).

The average grain yield of all variants (in both years of investigation and on both soil types) where the seed was treated by nodule bacteria *B. japonicum* is 3.64 t ha^{-1}. This is 11.86% less than the average grain yield of variants where the seeds were treated with bacteria *B. japonicum, A. brasilense and A. chroococcum* (4.13 t ha^{-1}).

Previous reports suggest an advantage of combined inoculation of *B. japonicum* with *A. brasilense* and *A. chroococcum*. Abdul Jabbar and Mohd Saud (2012) inoculated soybean seeds with *B. japonicum* and *A. brasilense* to obtain a higher grain yield of 27.26 g per plant. Phytohormones produced by *Azospirillum* promoted epidermal-cell differentiation in root hairs that increased the number of potential sites for rhizobial infection (Yahalom et al., 1991) leading to form more nodules (Andreeva et al., 1993).

Inoculation with mixed cultures of *B. japonicum*, *A. brasilense*, and *A. chroococcum* resulted in increased yields of soybean (Crossman and Hill, 1987; Herschkovitz et al., 2005). The ability of *Azotobacter* and *Azospirillum* to alter root morphology and plant growth rates has been related to the production of biologically active substances by these bacteria (Bashan and Levanony, 1990; Becking, 1992). *Azotobacter* and *Azospirillum* are active producers of phytohormones and vitamins of group B (Martinez-Toledo et al., 1991; Rodelas et al., 1993).

Average Number of Pods Per Plant

For both years of conducted research Humogley soil (Table 5) had the highest average number of pods per plant in variant 9 (seed treated with symbiotic bacteria *B. japonicum,* associative bacteria *A. brasilense* and non symbiotic bacteria *A. chroococcum* + reduced mineral nitrogen fertilization by 70%). All other variants of our data had lower average number of pods per plant with a statistically high significance ($p < 0.01$). Seed treatment with *B. japonicum* resulted in the highest average number of pods per plant in variant 5 (reduced mineral nitrogen fertilization by 70%). In the first year of research there was not a statistically significant difference between variant 1 – control (untreated seed + mineral nitrogen fertilization based on soil analysis) and variant 2 (seed treated only with symbiotic bacteria *B. japonicum* + mineral nitrogen fertilization based on soil analysis).

Eutric Cambisols soils had the highest average number of pods per plant in the first year of investigation in variant 8 (seed treated with symbiotic bacteria *B. japonicum,* associative bacteria *A. brasilense* and non symbiotic bacteria *A. chroococcum* + reduced mineral nitrogen fertilization by 50%). All other variants of research had lower average number of pods per plant with a statistical significance ($p<0.01$). Seed treatment with only the bacterium *B. japonicum* resulted in the highest average number of pods per plant in variant 4 (seed treated with symbiotic bacteria *B. japonicum* + reduced mineral nitrogen fertilization by 50%), but there were no statistically significant differences ($p>0.01$) between the listed variants and variant 7 (seed treated with symbiotic bacteria *B. japonicum,* associative bacteria *A. brasilense* and non symbiotic bacteria *A. chroococcum* + reduced mineral nitrogen fertilization by 30%). During the second year of investigation the largest average number of pods per plant was in variant 7 (seed treated with symbiotic bacteria *B. japonicum,* associative bacteria *A. brasilense* and non symbiotic bacteria *A. chroococcum* + reduced mineral nitrogen fertilization by 30%). All other variants achieved a statistically significant ($p<0.01$) lower average number of pods per plant.

Average number of pods per plant of all variants (in both years of investigation and on both types of the soil) where the seed was treated with *B. japonicum* is 24.94. This is 17.20% less than the average number of pods per plant of variants in which the seeds were treated with *B. japonicum, A. brasilense* and *A. chroococcum* (30.12). Soybean grain yield had a significant positive correlation with the number of pods per plant ($r = 0961$, $p<0.01$).

Our results concur with Patil et al. (2004) and Ara et al. (2009). Combined inoculation of soybean seed with *B. japonicum* and *A. brasilense* results in higher average number of pods per plant, 16.37, when compared to seed inoculation with only *B. japonicum.*

The results of Ara et al. (2009) studies suggest that higher average number of pods per plant was obtained when seeds were inoculated with *B. japonicum* and *A. chroococcum* - 14.52% higher average number of pods per plant when compared to seeds inoculated with only *B. japonicum.*

The beneficial effects of *Azotobacter* and *Azospirillum* on plants are mainly attributed to improvements in root development, an increase in the rate of water and mineral uptake by roots, the displacement of fungi and plant pathogenic bacteria, and biological nitrogen fixation (Okon and Itzigsohn, 1995).

Average Number of Grains Per Pod

In both years of our investigation the highest average number of grains per pod on soil type Humogley (Table 6) was observed in variant 9 (seed treated with symbiotic bacteria *B. japonicum,* associative bacteria *A. brasilense* and non symbiotic bacteria *A. chroococcum* + reduced mineral nitrogen fertilization by 70%). All other variants of research achieved a statistically significant ($p < 0.01$) lower average number of grains per pod. In variants inoculated with only *B. japonicum*, the highest average number of grains per pod was achieved with a reduction in mineral nitrogen fertilization by 50%. However, in 2011 there was no statistically significant difference ($p > 0.01$) in relation to the reduction of mineral nitrogen fertilization by 70%.

In the first year of investigation soil type Eutric Cambisols had the highest average number of grains per pod in variant 8 (seed treated with symbiotic bacteria *B. japonicum,* associative bacteria *A. brasilense* and non symbiotic bacteria *A. chroococcum* + reduced mineral nitrogen fertilization by 50%), while all other variants achieved a statistically significant ($p < 0.01$) lower average number of grains per pod. The highest average number of grains per pod was obtained by the reduction of mineral nitrogen fertilization of 50% in variants inoculated only with *B. japonicum.*

In the second year of investigation the highest average number of grains per pod was observed in variant 7 (seed treated with symbiotic bacteria *B. japonicum,* associative bacteria *A. brasilense* and non symbiotic bacteria *A. chro-ococcum* + reduced mineral nitrogen fertilization by 30%). All other variants had lower average grains per pod and were statistically significant ($p < 0.01$). In the variants inoculated only with *B. japonicum* the highest average number of grains per pod was obtained by the reduction of mineral nitrogen fertilization by 30%.

Differences between the variant 1 – control (untreated seed + mineral nitrogen fertilization based on soil analysis) and variant 2 (seed treated with symbiotic bacteria *B. japonicum* + mineral nitrogen fertilization based on soil analysis) were not statistically significant ($p>0.01$) on Humogley during the first year of investigation and Eutric Cambisols in the second year of investigation. In the second year of investigation on Humogley and the first year on Eutric Cambisols, there were no statistically significant differences ($p>0.01$) between the listed variants and variant 6 (seed treated with symbiotic bacteria *B. japonicum,* associative bacteria *A. brasilense* and non symbiotic bacteria *A. chroococcum* + mineral nitrogen fertilization based on soil analysis). This suggests that high levels of nitrogen have an inhibitory effect on average number of grains per pod.

Average number of grains per pod in all test types (in both years and on both types of soil) where the seed was treated by the nodule bacterium *B. japonicum* is 2.53. This is 13.36% less than the average number of grains per pod of variants in which the seeds were treated with *B. japonicum*, *A. brasilense* and *A. chroococcum* (2.92).

Average number of grains per pod has a significant positive correlation with the number of pods per plant ($r=0.939$; $p<0.01$). Similar results have been observed where Rodelas et al. (1999) demonstrated more elements of yield in faba beans (*Vicia faba* L.) when they are co-inoculated with *Rhizobium*, *Azotobacter* and *Azospirillum*.

Average Number and Dry Weight of Nodules

On soil type Humogley, for both years of investigation, the highest average number of nodules (Table 7) and average dry weight of nodules (Table 8) was achieved in variant 9 (seed treated with symbiotic bacteria *B. japonicum,* associative bacteria *A. brasilense* and non symbiotic bacteria *A. chroococcum* + reduced mineral nitrogen fertilization by 70%).

All other variants of research achieved a statistically significant ($p<0.01$) lower average number of nodules and average dry weight of nodules. When seeds are treated only with *B. japonicum*, the highest average number of nodules and average dry weight of nodules was achieved in variant 5 (reduced mineral nitrogen fertilization by 70%).

On soil type Eutric Cambisols in first year of investigation the highest average number of nodules and average dry weight of nodules was achieved in variant 8 (seed treated with symbiotic bacteria *B. japonicum,* associative bacteria *A. brasilense* and non symbiotic bacteria *A. chroococcum* + reduced mineral nitrogen fertilization by 50%). All other variants of investigation achieved a statistically significant ($p<0.01$) lower average number of nodules and average dry weight of nodules.

When it comes to seed treatment only by bacterium *B. japonicum*, the highest average number of nodules and average dry weight of nodules was achieved in variant 4 (reduced mineral nitrogen fertilization by 50%).

During the second year of investigation the highest average number of nodules and average dry weight of nodules was achieved in variant 7 (seed treated with symbiotic bacteria *B. japonicum,* associative bacteria *A. brasilense* and non symbiotic bacteria *A. chroococcum* + reduced mineral nitrogen fertilization by 30%). All other variants had a statistical significance ($p<0.01$) lower average number of nodules and average dry weight of nodules. In the variants inoculated only with bacteria *B. japonicum,* the highest average number of nodules and average dry weight of nodules was also obtained by the reduction of mineral nitrogen fertilization by 30%.

The average number of nodules of all variants (in both years of investigation and on both types of soil) in which the seed was treated with *B. japonicum* is 29.36 and 33.04% less than the average number of nodules of variants in which the seed was treated with *B. japonicum*, *A. brasilense* and *A. chroococcum* (43.85 nodules). Additionally, when seeds were treated only with *B. japonicum* average dry weight of nodules of all variants (on both years and in both types of soil) is 0.248 g and 30.14% lower than the average dry weight of nodules when the seed was treated with *B. japonicum, A. brasilense* and *A. chroococcum* (0.355 g).

Other studies with *Azospirillum*, *Azotobacter* and *Rhizobium* inoculum combinations have resulted in increased nodulation (Iruthayathas et al., 1983; Plazinski et al., 1984; Milić et al., 2002). Mixed cultures of *Azotobacter* spp. and *Rhizobium* strains when used as inoculants for several legumes caused the formation of increased numbers of root nodules when compared with plants inoculated with rhizobia alone (Burns et al., 1981; Poi et al., 1989; Remans et al., 2008).

Moreover, dual inoculation of soybean with *B. japonicum* and *A. brasilense* gave a significantly higher proportion of nodules attached to the main root, increased number of the most active nodules, and increased 23% of nitrogen content of soybean plants over *B. japonicum* single inoculated plant (Groppa et al., 1998).

Under dual inoculation (*Azospirillum* and *Rhizobium*), the weight of nodules increased to 50 mg per plant, compared with the control which registered only 8 mg per plant (Abdul Jabar and Mohd Saud, 2012).

Average Grain Protein Content

On soil type Humogley, for both years of investigation, the highest average grain protein content (Table 9) was observed in variant 9 (seed treated with symbiotic bacteria *B. japonicum,* associative bacteria *A. brasilense* and non symbiotic bacteria *A. chroococcum* + reduced mineral nitrogen fertilization by 70%).

There was no statistically significant difference between listed variant and variant 8 (seed treated with symbiotic bacteria *B. japonicum,* associative bacteria *A. brasilense* and non symbiotic bacteria *A. chroococcum* + reduced mineral nitrogen fertilization by 50%), variant 4 (seed treated with symbiotic bacteria *B. japonicum* + reduced mineral nitrogen fertilization by 50%) and variant 5 (seed treated with symbiotic bacteria *B. japonicum* + reduced mineral nitrogen fertilization by 70%). All other variants of research achieved a statistically significant ($p<0.01$) lower average grain protein content.

In the first year of investigation on soil type Eutric Cambisols the highest average grain protein content was observed in variant 8 (seed treated with symbiotic bacteria *B. japonicum,* associative bacteria *A. brasilense* and non symbiotic bacteria *A. chroococcum* + reduced mineral nitrogen fertilization by 50%), but there was no statistically significant difference between listed variant and variant 9 (seed treated with symbiotic bacteria *B. japonicum,* associative bacteria *A. brasilense* and non symbiotic bacteria *A. chroococcum* + reduced mineral nitrogen fertilization by 70%), variant 7 (seed treated with symbiotic bacteria *B. japonicum,* associative bacteria *A. brasilense* and non symbiotic bacteria *A. chroococcum* + reduced mineral nitrogen fertilization by 30%) and variant 4 (seed treated with symbiotic bacteria *B. japonicum* + reduced mineral nitrogen fertilization by 50%).

All other variants of research achieved a statistically significant ($p<0.01$) lower average grain protein content.

Table 4. Soybean grain yield (t ha^{-1}) on Humogley and Eutric Cambisol in 2011 and 2012 year

Variants				Type of soil / Year						*Average*
				Humogley			Eutric Cambisols			
				2011	2012	*Average*	2011	2012	*Average*	
Control (I) (1)				3.76	3.19	*3.48*	3.10	3.01	*3.06*	*3.27*
Fertilization	I	Treating seeds with bacteria	A (2)	3.78	3.22	*3.50*	3.18	3.09	*3.14*	*3.32*
			B (6)	4.19	3.69	*3.94*	3.26	3.15	*3.21*	*3.58*
	II		A (3)	4.26	3.80	*4.03*	3.51	3.43	*3.47*	*3.75*
			B (7)	4.61	4.18	*4.40*	3.99	3.89	*3.94*	*4.17*
	III		A (4)	4.42	3.84	*4.13*	3.72	3.39	*3.56*	*3.85*
			B (8)	5.03	4.66	*4.85*	4.06	3.76	*3.91*	*4.38*
	IV		A (5)	4.17	3.67	*3.92*	3.44	3.19	*3.37*	*3.65*
			B (9)	5.16	4.97	*5.07*	3.80	3.41	*3.67*	*4.37*
$LSD_{0.05}$				0.214	0.166	0.189	0.135	0.140	0.139	0.163
$LSD_{0.01}$				0.381	0.308	0.344	0.248	0.256	0.251	0.310

Fertilization: I – mineral nitrogen fertilization based on soil analysis; II – reduced mineral nitrogen fertilization by 30%; III – reduced mineral nitrogen fertilization by 50%; IV – reduced mineral nitrogen fertilization by 70%. Treating seeds with bacteria: A – seed treated with symbiotic bacteria *B. japonicum*; B – seed treated with symbiotic bacteria *B. japonicum*, associative bacteria *A. brasilense* and non symbiotic bacteria *A. chroococcum*. (1), (2), (3), (4), (5), (6), (7), (8), (9) – Listed variants in text.

Table 5. Average number of pods per plant on Humogley and Eutric Cambisol in 2011 and 2012 year

Variants				Type of soil / Year						*Average*
				Humogley			Eutric Cambisols			
				2011	2012	*Average*	2011	2012	*Average*	
Control (I) (1)				21.31	19.93	*20.62*	22.09	23.05	*22.57*	*21.60*
Fertilization	I	Treating seeds with bacteria	A (2)	21.45	20.53	*20.99*	21.48	21.07	*21.27*	*21.13*
			B (6)	26.13	24.18	*25.16*	24.03	24.97	*24.50*	*24.83*
	II		A (3)	27.11	26.13	*26.62*	22.97	22.91	*22.94*	*24.78*
			B (7)	30.82	30.05	*30.44*	26.31	28.91	*27.61*	*29.03*
	III		A (4)	29.14	29.05	*29.10*	26.96	22.16	*24.56*	*26.83*
			B (8)	37.08	36.11	*36.60*	31.55	26.04	*28.80*	*32.70*
	IV		A (5)	33.45	30.41	*31.93*	24.07	20.18	*22.13*	*27.03*
			B (9)	43.65	39.04	*41.35*	28.03	24.90	*26.47*	*33.91*
$LSD_{0.05}$				0.360	0.319	0.331	0.314	0.294	0.299	0.318
$LSD_{0.01}$				0.635	0.572	0.596	0.582	0.559	0.565	0.588

Fertilization: I – mineral nitrogen fertilization based on soil analysis; II – reduced mineral nitrogen fertilization by 30%; III – reduced mineral nitrogen fertilization by 50%; IV – reduced mineral nitrogen fertilization by 70%. Treating seeds with bacteria: A – seed treated with symbiotic bacteria *B. japonicum*; B – seed treated with symbiotic bacteria *B. japonicum,* associative bacteria *A. brasilense* and non symbiotic bacteria *A. chroococcum.* (1), (2), (3), (4), (5), (6), (7), (8), (9) – Listed variants in text.

Table 6. Average number of grains per pod on Humogley and Eutric Cambisol in 2011 and 2012 year

Variants				Type of soil / Year						*Average*
				Humogley			Eutric Cambisols			
				2011	2012	*Average*	2011	2012	*Average*	
Control (I) (1)				2.04	2.12	*2.08*	2.14	2.07	*2.11*	*2.10*
Fertilization	I	Treating seeds with bacteria	A (2)	2.11	2.11	*2.11*	2.09	2.10	*2.10*	*2.11*
			B (6)	2.50	2.19	*2.35*	2.25	2.60	*2.43*	*2.39*
	II		A (3)	2.71	2.38	*2.55*	2.44	2.53	*2.49*	*2.52*
			B (7)	3.20	3.12	*3.16*	2.75	2.87	*2.81*	*2.99*
	III		A (4)	3.08	3.01	*3.05*	2.83	2.39	*2.61*	*2.83*
			B (8)	3.36	3.29	*3.33*	3.06	2.70	*2.88*	*3.11*
	IV		A (5)	2.98	2.80	*2.89*	2.50	2.28	*2.39*	*2.64*
			B (9)	3.96	3.51	*3.74*	2.79	2.49	*2.64*	*3.19*
$LSD_{0.05}$				0.084	0.072	0.077	0.068	0.064	0.067	0.072
$LSD_{0.01}$				0.157	0.139	0.146	0.125	0.118	0.120	0.133

Fertilization: I – mineral nitrogen fertilization based on soil analysis; II – reduced mineral nitrogen fertilization by 30%; III – reduced mineral nitrogen fertilization by 50%; IV – reduced mineral nitrogen fertilization by 70%. Treating seeds with bacteria: A – seed treated with symbiotic bacteria *B. japonicum*; B – seed treated with symbiotic bacteria *B. japonicum,* associative bacteria *A. brasilense* and non symbiotic bacteria *A. chroococcum.* (1), (2), (3), (4), (5), (6), (7), (8), (9) – Listed variants in text.

Table 7. Average number of nodules on Humogley and Eutric Cambisol in 2011 and 2012 year

Variants				Type of soil / Year						Average
				Humogley			Eutric Cambisols			
				2011	2012	*Average*	2011	2012	*Average*	
Control (I) (1)				9.08	7.14	*8.11*	6.82	7.01	*6.92*	*7.52*
Fertilization	I	Treating seeds with bacteria	A (2)	19.65	18.94	*19.30*	19.07	17.02	*18.05*	*18.68*
			B (6)	26.57	25.30	*25.94*	24.11	24.97	*24.54*	*25.24*
	II		A (3)	27.14	20.56	*23.85*	30.05	35.01	*32.53*	*28.19*
			B (7)	51.14	45.18	*48.16*	47.03	47.12	*47.08*	*47.62*
	III		A (4)	39.50	31.10	*35.30*	38.10	30.05	*34.08*	*34.69*
			B (8)	59.65	52.03	*55.84*	53.05	41.62	*47.34*	*51.59*
	IV		A (5)	48.16	39.63	*43.90*	32.11	23.55	*27.83*	*35.87*
			B (9)	64.55	57.10	*60.83*	49.68	32.47	*41.08*	*50.96*
$LSD_{0.05}$				0.793	0.784	0.788	0.696	0.703	0.701	0.740
$LSD_{0.01}$				1.514	1.501	1.504	1.412	1.398	1.402	1.447

Fertilization: I – mineral nitrogen fertilization based on soil analysis; II – reduced mineral nitrogen fertilization by 30%; III – reduced mineral nitrogen fertilization by 50%; IV – reduced mineral nitrogen fertilization by 70%. Treating seeds with bacteria: A – seed treated with symbiotic bacteria *B. japonicum*; B – seed treated with symbiotic bacteria *B. japonicum,* associative bacteria *A. brasilense* and non symbiotic bacteria *A. chroococcum.* (1), (2), (3), (4), (5), (6), (7), (8), (9) – Listed variants in text.

Table 8. Average dry weight of nodules (g) on Humogley and Eutric Cambisol in 2011 and 2012 year

Variants				Type of soil / Year						*Average*
				Humogley			Eutric Cambisols			
				2011	2012	*Average*	2011	2012	*Average*	
Control (I) (1)				0.042	0.027	*0.035*	0.020	0.024	*0.022*	*0.029*
Fertilization	I	Treating seeds with bacteria	A (2)	0.191	0.154	*0.173*	0.184	0.179	*0.182*	*0.178*
			B (6)	0.238	0.219	*0.229*	0.211	0.209	*0.210*	*0.220*
	II		A (3)	0.251	0.180	*0.216*	0.229	0.270	*0.250*	*0.233*
			B (7)	0.419	0.353	*0.386*	0.365	0.371	*0.368*	*0.377*
	III		A (4)	0.320	0.250	*0.285*	0.309	0.234	*0.272*	*0.279*
			B (8)	0.480	0.411	*0.446*	0.448	0.322	*0.385*	*0.416*
	IV		A (5)	0.387	0.324	*0.356*	0.275	0.215	*0.245*	*0.301*
			B (9)	0.527	0.461	*0.494*	0.389	0.247	*0.318*	*0.406*
$LSD_{0.05}$				0.029	0.030	0.030	0.024	0.022	0.024	0.028
$LSD_{0.01}$				0.057	0.058	0.059	0.041	0.038	0.040	0.052

Fertilization: I – mineral nitrogen fertilization based on soil analysis; II – reduced mineral nitrogen fertilization by 30%; III – reduced mineral nitrogen fertilization by 50%; IV – reduced mineral nitrogen fertilization by 70%. Treating seeds with bacteria: A – seed treated with symbiotic bacteria *B. japonicum*; B – seed treated with symbiotic bacteria *B. japonicum,* associative bacteria *A. brasilense* and non symbiotic bacteria *A. chroococcum.* (1), (2), (3), (4), (5), (6), (7), (8), (9) – Listed variants in text.

Table 9. Average grain protein content (%) on Humogley and Eutric Cambisol in 2011 and 2012 year

Variants				Type of soil / Year						*Average*
				Humogley			Eutric Cambisols			
				2011	2012	*Average*	2011	2012	*Average*	
Control (I) (1)				30.824	31.109	*30.967*	30.094	30.015	*30.055*	*30.511*
Fertilization	I	Treating seeds with bacteria	A (2)	30.915	31.014	*30.965*	30.085	32.515	*31.300*	*31.133*
			B (6)	31.015	32.008	*31.512*	32.256	33.218	*32.737*	*32.125*
	II		A (3)	31.943	32.146	*32.045*	35.014	35.964	*35.489*	*33.767*
			B (7)	34.154	35.292	*34.723*	36.918	37.092	*37.005*	*35.864*
	III		A (4)	35.805	35.940	*35.873*	36.201	35.892	*36.047*	*35.960*
			B (8)	37.011	37.463	*37.237*	37.513	36.516	*37.015*	*37.126*
	IV		A (5)	36.550	36.017	*36.284*	34.210	34.017	*34.114*	*35.199*
			B (9)	38.064	37.915	*37.990*	36.404	35.983	*36.194*	*37.092*
$LSD_{0.05}$				1.470	1.105	1.130	0.969	0.887	0.960	1.071
$LSD_{0.01}$				2.260	1.983	2.119	1.719	1.660	1.711	1.903

Fertilization: I – mineral nitrogen fertilization based on soil analysis; II – reduced mineral nitrogen fertilization by 30%; III – reduced mineral nitrogen fertilization by 50%; IV – reduced mineral nitrogen fertilization by 70%. Treating seeds with bacteria: A – seed treated with symbiotic bacteria *B. japonicum*; B – seed treated with symbiotic bacteria *B. japonicum,* associative bacteria *A. brasilense* and non symbiotic bacteria *A. chroococcum.* (1), (2), (3), (4), (5), (6), (7), (8), (9) – Listed variants in text.

During the second year of investigation on soil type Eutric Cambisols the highest average grain protein content was observed in variant 7 (seed treated with symbiotic bacteria *B. japonicum,* associative bacteria *A. brasilense* and non symbiotic bacteria *A. chroococcum* + reduced mineral nitrogen fertilization by 30%), but there was no statistically significant difference between listed variant and variant 8 (seed treated with symbiotic bacteria *B. japonicum,* associative bacteria *A. brasilense* and non symbiotic bacteria *A. chroococcum* + reduced mineral nitrogen fertilization by 50%), variant 9 (seed treated with symbiotic bacteria *B. japonicum,* associative bacteria *A. brasilense* and non symbiotic bacteria *A. chroococcum* + reduced mineral nitrogen fertilization by 70%) including the variants inoculated only with bacteria *B. japonicum* (variants 3, 4 - reduced mineral nitrogen fertilization by 30% and 50%).

The average grain protein contents of all variants (in both years of investigation and on both types of soil) in which the seed was treated with bacteria *B. japonicum* is 34.01% and 4.33% less than the average grain protein contents of variants where the seed was treated with bacteria *B. japonicum*, *A. brasilense* and *A. chroococcum* (35.55%).

The average grain protein content of control (untreated seed + mineral nitrogen fertilization based on soil analysis) was 30.51% and 10.29% less than the average grain protein contents in variants where the seed was treated by nodule bacteria *B. japonicum,* or 14.18% less than the average grain protein contents in variants where the seed was treated with bacteria *B. japonicum*, *A. brasilense* and *A. chroococcum.*

Research results are in accordance with the results of many authors that claim that inoculation of soybean seed with nitrogen - fixing bacteria increased grain protein content (Dashti et al., 1996; Egamberdiyeva et al., 2004; Aung et al., 2013; Berquó Marks et al., 2013).

CONCLUSION

Based on these analyzes, and results obtained, we can perform following conclusions:

- Variants where the seed was inoculated with nitrogen-fixing bacteria, only with *B. japonicum* or in combination with *A. brasilense* and *A. chroococcum* achieved significantly higher soybean grain yield, average number of pods per plant, average number of grains per pod, average number of nodules, average dry weight of nodules and average grain protein content compared to control.

- For almost all investigated parameters (except average grain protein content) variants where soybean seeds were inoculated with symbiotic bacteria *B. japonicum*, associative bacteria *A. brasilense* and non symbiotic nitrogen fixing bacteria *A. chroococcum* achieved statistically significant higher average values of investigated elements of yield, unlike the variants where the seeds were inoculated only with symbiotic bacteria *B. japonicum*.
- The soil with better chemical and microbiological properties – Humogley, had the highest average values of the investigated parameters achieved by the reduction of nitrogen fertilization by 70%. However, in some variants, there were no statistically significant differences in comparison with variants where nitrogen fertilization was reduced by 50% (second year of investigation).
- The soil type with poorer chemical and microbiological properties - Eutric Cambisols, had the highest average values of the investigated parameters achieved by the reduction of nitrogen fertilization by 50%. However, in some variants, there were no statistically significant differences in comparison with variants where nitrogen fertilization was reduced by 30%. Especially, during the second year of investigation were recorded bad weather conditions for soybean production, primarily because of the lack of rainfall.

Reviewed by Ph.D. Aleksandra Sudarić, scientific consultant Chairman of Department for breeding and genetics of industrial plants. Institute of Agriculture in Osijek, Južno predgrađe 17, 31 000 Osijek, Republic of Croatia

References

Abdul Jabar, B. K. and Mohd Saud, H. (2012): Effects of molybdenum on biological nitrogen fixation by combination of rhizobium and azospirillum in soybean under drip irrigation system. *Int. J. Life Sc. Bt. and Pharm. Res.*, 1(2):63-77.

Anandham, R., Sridar, R., Nalayini, P., Poonguzhali, S., Madhaiyan, M. and Tongmin, S. (2007): Potential for plant growth promotion in groundnut

(*Arachis hypogaea* L.) cv. ALR-2 by co-inoculation of sulfuroxidizing bacteria and *Rhizobium, Microbiol. Res.* Vol. 162, pp. 139-153.

Andreeva, I., Redkina, T. and Izmailov, S. (1993): The involvement of indole-acetic acid in the stimulation of *Rhizobium*-legume symbiosis by *Azospirillum brasilense. Russ. J. Plant Physiol.*, 40:780-789.

Ara, N., Rokonuzzaman, M. and Hasan, M. N. (2009): Effect of *Bradyrhizobium* and *Azotobacter* on growth and yield of mungbean varieties. *J. Bangladesh Agril. Univ.* 7(1):7–13.

Aung, T. T., Tittabutr, P., Boonkerd, N., Herridge, D., and Teaumroong, N. 2013): Co-inoculation effects of *Bradyrhizobium japonicum* and *Azospirillum* sp. on competitive nodulation and rhizosphere eubacterial community structures of soybean under rhizobia-established soil conditions. *African Journal of Biotechnology*, 12(20):2850-2862.

Babeva, I. P. and Zenova, A. R. (1989): *Biologir. Počv., izd.,* Moskovskogov univerziteta, Moskva.

Bashan, Y., Harrison, K. and Whitmoyer, R. (1990): Enhanced growth of wheat and soybean plants inoculated with *Azospirillum brasilense* is not necessarily due to general enhancement of mineral uptake. *Appl. Environ. Microbiol.*, 56:769–775.

Bashan, Y. and Levanony, H. (1990): Current status of *Azospirillum* inoculation technology: *Azospirillum* as a challenge for agriculture. *Can. J. Microbiol.,* 67:1317–1324.

Becking, J. H.: The family *Azotobacteraceae.* In: Balows, A., Trüper, H. G., Dworkin, M., Harder, W., Schleifer, K. H. (eds.) *The prokaryotes*, 2nd edn. *A handbook on the biology of bacteria: ecophysiology, isolation, identification, applications*, vol. 4. Springer, Berlin Heidelberg New York, pp. 3144–3170, 1992.

Berquó Marks, B., Megías, M., Nogueira, M. A., and Hungria, M. (2013): Biotechnological potential of rhizobial metabolites to enhance the performance of *Bradyrhizobium* spp. and *Azospirillum brasilense* inoculants with soybean and maize. *AMB Express*, 3:21-22.

Bray, C. M. (1983): Interconversions of Nitrogen. In: *Nitrogen Metabolism in Plants,* ed. Longman Group Limited, British Library, 13:1-28.

Brown, M. E. (1974): Seed and root bacterization. *Ann. Rev. Phytopathol.*, 12: 181–197.

Burdman, S., Vedder, D., German, M., Itzigsohn, R., Kigel, J., Jurkevitch, E., and Okon, Y.: Legume crop yield promotion by inoculation with *Azospirillum.* In: C. Elmerich, A. Kondorosi, W. E. Newton. *Biological nitrogen fixation for the 21th century* (pp. 609-612). Netherlands. Kluwer Academic Publishers, 1998.

Burns, T. A. I. R., Bishop, P. and Israel, D. W. (1981): Enhanced nodulation of leguminous plant roots by mixed cultures of *Azotobacter vinelandii* and *Rhizobium. Plant and Soil,* 62:399-412.

Burns, R. G.: Enumeration, Survival, and Beneficial Activities of Microorganisms. Introduced into Soil. 145-164. In: Huang, P. M., et al., (eds.) *Environmental impact of soil component interaction, Metal, Other Inorganics, and Microbial Activities*, CRC Pres. Inc. London, 1995.

Crossman, S., Hill, W. (1987): Inoculation of sweet potato with *Azospirillum. Horti. Sci.*, 22(3): 420-422.

Cvijanović, G., Milošević, N. and Jarak, M. (2007): The importance of diazotrophs as biofertilisers in the maize and soybean production. - *Genetika*, 39(3):395-404.

Dashti, N., Zhang, F., Hynes, R., and Smith, D. L. (1996): Application of plant growth-promoting rhizobacteria to soybean (*Glycine max* [L.] Merr.) increases protein and dry matter yield under short-season conditions. *Plant and Soil* 188(1):33-41.

Dobbelaere, S. and Okon, Y.: The plant growth promoting effect and plant responses. In: C. Elmerich, W. E. Newton. *Associative and endophytic nitrogen fixing bacteria and Cyanobacterial associations*, Netherlands, Kluwer Academic Publishers, pp. 1-26, 2007.

Egamberdiyeva, D., Qarshieva, D. and Davranov, K. (2004): Growth and yield of soybean varieties inoculated with *Bradyrhizobium* spp. In; N-deficient calcareous soils. *Biology and Fertility of Soils*, 40(2):144-146.

Ergorov, V. J. (1985): Symbiotic nitrogen fixation in cultivated podzolic soils of the Kola Peninsula. *Microbiology*, 54(3):388-391.

Essa, T. A. and Dulaimi, H. M. (1985): Nodulation response of soybean to nitrogen application and inoculation. *Iraqi Journal of Agricultural Sciences, „Zanco“,* 3(3):27-35.

Groppa, M. D., Zawoznik, M. S. and Tomaro, M. L. (1998): Effect of co-inoculation with *Bradyrhizobium japonicum* and *Azospirillum brasilense* on soybean plants. *Eur. J. Soil Biol.*, 34(2):75-80.

Hardy, R. W. F. and Eaglesham, A. R. J.: Ecology and agricultural applications of nitrogenfixing systems overview In: *Nitrogen Fitation. Fundametals and Applications* (eds Tichonovich et. al): Currend Plant Science Academic Publish Boston. London, 1995.

Herschkovitz, Y., Lerner, A., Davidov, Y., Okon, Y., and Jurkevitch, E. (2005): *Azospirillum brasilense* does not affect population structure of specific rhizobacterial communities of inoculated maize (*Zea mays*). *Environ. Microbiol.*, 7(11):1847-1852.

Hungria, M. A. and Bihrer, T. R. J. (2000): Variability of nodulation and nitrogen fixation capacity among soybean cultivar. *Biol. Fertil. Soils,* 31:45-52.

Iruthayathas, E. E., Gunasecran, S. and Vlassak, K. K. (1983): Effect of combined inoculation soybean. *Scientia Hort.*, 20:231-240.

Jagnow, G. (1987): Inoculation of cereal crops and forage grasses with nitrogen fixing rhizosphere bacteria: possible causes of success and failure with regard to yield response – a review. *Z Pflanzenernaehr Bodenk.*, 150: 361–368.

Kristek, B. S. (2001): The effectiveness of inoculation, nitrogen fertilization and molybdenum on yield of field pea. *Doctoral dissertation*. Faculty of Agronomy. University of J.J. Strossmayer, Osijek, Croatia.

Lin, W., Okon, Y. and Hardy, R. W. (1983): Enhanced mineral uptake by *Zea mays* and *Sorghum bicolor* roots inoculated with *Azospirillum brasilense. Appl. Environ. Microbiol.*, 45:1775–1779.

Marre, E.: *Regulation of cell membrane activities in plants*. Marre, E. and Ceferni, O. (Eds.), North Holland, Amsterdam, 1976.

Martinez-Toledo, M. V., Salmerón, V. and González-López, J. (1991): Biological characteristics of *Azotobacter* in natural environments. *Trends Soil Sci.,* 1:15–23.

Milić, V., Mrkovački, N., Popović, M., and Malenčić, Đ. (2002): Nodule efficiency of three soybean genotypes inoculated by different methods. *Rostlinná Výroba*, 48(8):356-360.

Mrkovački, N., Saric, Z., Saric, R. M., and Milic, V. (1992): Symbiotic effectiveness of some soya bean genotypes. *Mikobiologija,* 29:1-26.

Murty, M. G. and Ladha, J. K. (1988): Influence of *Azospirillum* inoculation on the mineral uptake and growth of rice under hydroponic conditions. *Plant Soil*, 108:281–285.

Okon, Y. (1985): *Azospirillum* as potential inoculants for agriculture", *Trends Biotechnol.,* Vol. 3, pp. 223-228.

Okon, Y. and Kapulnik, Y. (1986): Development and function of *Azospirillum* inoculated roots, *Plant and Soil*, Vol. 90, pp. 3-16.

Okon, Y. and Labandera-González, C. (1994): Agronomic applications of *Azospirillum*: an evaluation of 20 years' worldwide field inoculation. *Soil Biol. Biochem.,* 26: 1591–1601.

Okon, Y. and Itzigsohn, R. (1995): The development of *Azospirillum* as a commercial inoculant for improving crop yields. *Biotechnol. Adv.*, 13: 415–424.

Pacovsky, R. S., Paul, E. A. and Bethlenfalvay, G. J. (1985): Nutrition of sorghum plants with nitrogen or inoculated with *Azospirillum brasilense*. *Plant Soil,* 85:145–148.

Patil, A. B., Naik, N. M., Jones, P., Nirmalnath, J., and Alagawadi, A. R. (2004): Effect of Combined Inoculations of Plant Growth Promoting Rhizobacteria (PGPR) on the Growth and Productivity of Soybean (Glycine max L.). *Karnataka J. Agric. Sci*., 17(4):731-735.

Plazinski, J., Gartner, E., Jahnke, R., and Rolfe, E. B. G.: Effect of *Azospirillum* strains on *Rhizobium* – legume symbiosis. In: Vegger, C., Newton, W. E. (eds.): *Advances in nitrogen fixation research*. Martinus Nijhoff Dv. W. Junk, Pudoc, Wageningen, The Hague, 1984.

Poi, S. C., Ghosh, G. and Kabi, M. C. (1989): Response of chick pea to combined inoculation with *Rhizobium* phosphobacteria and mycorrhizal organisms. *Zentralbl. Mikrobiolog*., 144:249–253.

Remans, R., Beebe, S., Blair, M., Manrique, G., Tovar, E., Rao, I., Croonenborghs, A., Torres-Gutierrez, R., El-Howeity, M., and Michiels, J. (2008): Physiological and genetic analysis of root responsiveness to auxin-producing plant growth-promoting bacteria in common bean (*Phaseolus vulgaris* L.). *Plant Soil*, 302(1):149-161.

Rodelas, B., Salmerón, V., Martinez-Toledo, M. V., and González-López, J. (1993): Production of vitamins by *Azospirillum brasilense* in chemically-defined media. *Plant Soil,* 153:97–101.

Rodelas, B., González-López, J., Salmerón, V., Pozo, C., and Martínez-Toledo, M. V. (1996): Enhancement of nodulation, N_2-fixation and growth of faba bean (*Vicia faba* L.) by combined inoculation with *Rhizobium leguminosarum* bv. viceae and *Azospirillum brasilense*. *Symbiosis*, 21:175–186.

Rodelas, B., González-López, J., Martínez-Toledo, M. V., Pozo, C., and Salmerón, V. (1999): Influence of *Rhizobium/Azotobacter* and *Rhizobium/ Azospirillum* combined inoculation on mineral composition of faba bean (*Vicia faba* L.). *Biol. Fertil. Soils*., 29:165–169.

Sadowski, M. J., Tully, R. E., Cregan, P. B., and Keyser, H. H. (1987): Genetic diversity in *Bradyrhizobium japonicum* serogroup 123 and relation to genotype-specific nodulation of soybean. *Appl. Envir. Microbiol.* 53:264-269.

Thies, J. E., Singleton, P. W. and Bohlool, B. B. (1995): Phenology, growth, and yield of field – grown soybean and bush bean as a function of varying modes of N nutrition. *Soil Biology and Biochemistry,* 27(4-5):575-583.

Tien, T. M., Gaskins, M. H. and Hubbel, D. H. (1979): Plant growth substances produced by *Azospirillum brasilense* and their effect on the growth of pearl millet. *Applied and Environmental Microbiology*, 37: 1016-1024.

Triplett, E. W. and Sadowsky, M. J. (1992): Genetics of competition for nodulation of legumes. *Annu. Rev. Microbiol.*, 46:399-428.

Waterer, J. G., Vessey, J. K. and Raper, C. D. (1992): Stimulation of nodulation in field peas (*Pisum sativum*) by low concentrations of ammonium in hydroponic culture. *Physiologia Plantarum,* 86(2):215-220.

Yahalom, E., Okon, Y. and Dovrat, A. (1987): *Azospirillum* effects on susceptibility to *Rhizobium* nodulation and on nitrogen fixation of several forage legumes. *Can. J. Microbiol.*, 33:510–514.

Yahalom, E., Dovrat, A., Okon, Y., and Czosnek, H. (1991): Effect of inoculation with *Azospirillum brasilense* strain Cd and Rhizobium on the root morphology of burr medic (*Medicago polymorpha* L.). *Isr. J. Bot.,* 40(2): 155-164.

In: Nitrogen Fertilizer
Editor: Karlene Winslow

ISBN: 978-1-63321-585-6

Chapter 4

REVIEW OF ANALYSES FOR NITROGEN SUPPLY IN MAIZE AND WINTER WHEAT - CASE OF DEVELOPING COUNTRIES I.E. SLOVENIAN EXPERIENCES

Franc Bavec and Martina Bavec
University of Maribor, Faculty of Agriculture and Life Sciences, 2312Hoce/Maribor, Slovenia

ABSTRACT

In developing countries the supply of cash crops of nitrogen is usually based on unfounded rates and did not successfully change–with fertilization, based on N analyses and calculated nitrogen target values. In Slovenia, the first analyses of the sap test in winter wheat were done 30 years ago, but the soil Nmin analyses were not supported from university professors and even from the government until 2004. The soil Nmin sometimes reached more than 400 kg Nmin ha^{-1} up to 0.9 m soil depth, which was associated with nitrogen pollution of drinking water with over 50 mg $N0_3^-$ l^{-1}. In the late 1990s, three researchers (Bavec, Bavec and Briški) provided analyses of soil Nmin and established nitrogen target values despite objections. Analyses are involved for practical farm use with EU support as an agri-environmental scheme after establishing integrated field crop production standards. Slovenia is the only country that provided this measure at a national level in EU 27 in the period 2004-2014. The data from research and analyses of integrated crop production praxis with a defined system of soil Nmin analyses were done in this chapter. Presentation of Nmin data and some problems of introduction of soil Nmin analyses into the support system were analyzed and represented. The analyses of Nmin, based on 5075 soil samples, also showed that the timing of sampling was sometimes incorrect, but timely analyses showed appropriate levels of Nmin in the soil profiles. However, in the new Program of rural development for 2014-2020, the nitrogen fertilization based on the soil Nmin analyses, for the Slovenian Ministry of agriculture and the environment, is a system too complicated for money transfer for economically and environmentally acceptable N supply in field crop production, which complies to EU CAP and EU Nitrate directive. However, producers, professionals, non-governmental and governmental institutions need to renew calculations of economical benefits, impacts and environmental risks for better development and reasonable N supply in production of their main cash crop. Because of discrepancies among countries, the minimal N management standards due to production costs and negative environmental impacts needs to be regulated and supported around the world by the FAO and/or OECD.

INTRODUCTION

N supply based on analyses was one of the first topics of fertilization development in agronomy at the end of 20th century in developed countries, but their implication depends on societal, regional or national requirements associated with priorities of economical and environmental importance. Establishing the correct N fertilization and especially top dressing rates must be based on different factors such as crop genotype, expected yield and quality, available N in the soil, mineralization potential and plant needs during the growing period. European farmers usually meet the needs written in guidelines, but in investigation reports and papers we can find many pro et contra conclusions about appropriate analyses like total N, Nmic, N org, soil Nmin or NO_3-N analyses (as a main tool in USA), sap NO_3^- or total N plant tests, chlorophyll meter readings, canopy sensors, etc., and only a few of them are implemented into practice. Former traditional understanding and requirements were based on advised nitrogen rates without any analyses, where some leading professionals advised based on 'practical experiences' and this kind of practice is very difficult to transform into environmentally responsible field crop production. Furthermore in some developing countries we can find a discrepancy between scientific findings, fertilization knowledge transferred to farmers, and governmental supported and advised recommendations based on analyses. In some countries none of this type of support exists.

The first analyses of sap test in winter wheat in Slovenia were done 30 years ago. The soil Nmin analyses were not supported from university professors and even from government until 2004. The soil Nmin sometimes reached even more than 400 kg Nmin ha^{-1} up to 0.9 m soil depth, which was associated with nitrogen pollution of drinking water with over 50 mg $N0_3$- l^{-1}. Since the late 1990s, three researchers (Bavec, Bavec and Briški), provided analyses of soil Nmin and established nitrogen target values in spite of objections from leading researchers. Analyses are involved with practical farmer use with EU support as an agri-environmental scheme after establishing integrated field crop production guidelines in 2004, as the only country that provided this measure at a national level in EU 27 in the period 2004-2014.

In the EU, due to the Nitrate directive (Council Directive 91/676/EEC, Regulation (EC) No 1882/2003, Regulation (EC) No 1137/2008) concerning the protection of waters against pollution caused by nitrates from agricultural sources, the agricultural practice has to adapt nitrogen fertilization management.

The aim of this chapter is to analyze investigations of nitrogen fertilization in the case of maize and winter wheat in Slovenia and their implementation into the practical sector, with special attention on agronomical and environmental impacts.

Data of N Supply Based on Analyses

Slovenian Experience

Amongst progression of small and undeveloped countries compared to well-organized developed countries there exists a lot discrepancies. However in Slovenia, a few basic extensive studies exist.

In maize (*Zea mays* L.), N fertilization trials at loam-sandy soils at experimental circumstances at Maribor, Slovenia (36° 34 N, 15° 38 E latitude, 274 a.s.l., yearly precipitation 1000 mm) concluded (Bavec, 1992) that influence of nitrogen fertilization rates (55, 150, 225, 300 and 0 kg N ha^{-1} – control plot) on maize expressed a small influence on grain yield (R: up to 0.19). But maize yield and soil Nmin before sowing was firmly correlated (R: 0.42 - 0.63) and even more firmly (R: 0.55 – 0.82) at the growth stage of 7 to 9 leaves (BBCH 17 – 19). The correlation coefficient between Nmin and yield were very firm; if the soil was 157 kg Nmin ha^{-1} before sowing then the fertilization was (r= 0.41 – 0.83), or 274 kg Nmin ha^{-1} at BBCH 17-19 (r: 0.56-0.75). In the case of low soil Nmin before fertilization at sowing time the effects of added nitrogen had a firm correlation with grain yield. In this case the maximum yields were established up to a total of 325 kg N (Nmin plus N added with fertilizers) ha^{-1}. This means that at 73 to 79 kg Nmin ha^{-1} before sowing the yields were increased up to 150 kg added N ha^{-1}, in split rates 75 kg N ha^{-1} before sowing and 75 kg N ha^{-1} at BBCH 17-19 stage. Also based on soil NO_3-N, firm correlations with grain yield were found. It was the main research, and when we took additional data (Bavec, 2001) into account, the advised nitrogen rate for maize top dressing is Kg N ha^{-1} = 225 to 325 kg N ha^{-1} – kg Nmin ha^{-1} (from 0-0.9 m soil deep), and as optimal NO_3-N in the soil 21 mg NO_3-N to soil depth 0.3 m was advised (Annon., 2014) as an approximate level that we begin to consider additional agronomical N needs in crops and negative nitrogen impacts of additional fertilization with nitrogen. In general, additional fertilization in this case is not necessary. Additional results of trials measuring N uptake of the total dry mass of above ground plants showed an increase from 159 to 255 kg N ha^{-1} (Bavec, 1992). However, the effects of fertilization, nitrogen use efficiency (NUE) and N accumulation were conditioned by the yearly circumstances (and Nmin), hybrids, plant densities, method of sowing and numerous other interactions.

With the introduction of sweet maize (*Zea mays* L. *sacharata* Sturt.) into temperate climates, additional information concerning fertilization is required, especially for developing an organic production system. Field experiments were carried out in the Slovenian region, Styria, suitable for growing only early-maturity maize hybrids (FAO100–400), with the aim of determining the effects of nitrogen applied to different nitrogen target values (70, 120, 170 and 220 kg ha^{-1} N) on growth, yield, photosynthetic activity and soil mineral nitrogen (Nmin – NO_3-N and NH4-N) dynamics, as compared to the control. Nitrogen was applied as organic by-products (pumpkin cake and pig manure digestate) and mineral fertilizers (CAN 27 and ENTEC® 26). The major points were as follows: 1) pumpkin cake had a similar effect as that of mineral fertilizers, and gave a significantly higher total and marketable yields (14.476 and 11.619 t ha^{-1}, respectively), higher values of cob characteristics and plant mass than for pig manure digestate; 2) there were no significant differences in total and marketable yields among the target values of 120, 170 and 220 kg ha^{-1} N, but the calculated nitrogen target value, expressed as the peak of a regression curve for yield, was 170 kg ha^{-1} N. However, the data showed that high yields of organic sweet maize could be obtained using pumpkin cake as nitrogen fertilizer based on a target value 120 kg ha^{-1} N (rate of kg ha^{-1} N = 120 kg ha^{-1} N – soil Nmin till 0.9 m depth) at the time of sowing (Bavec et al., 2013).

Experiences with chlorophyll meter readings and plant sap tests in maize are discussable. When we compare four different N rates at four hybrids, the differences among readings and hybrids had the same variation. It a reason that we concluded (Bavec[a], 1994) that optimization for maize N fertilization based on chlorophyll meter readings must be evaluated for each genotype, which is not a solution for practical applications. The same has happened with NO_3^- sap tests: there were no clear correlations between NO_3^- levels and grain yield (Bavec[b], 1994). In spite of our results recommendations were made (Leskošek, 1994), but after discussion they were never recommended again. (ii) In winter wheat the first advice came from analyses of sap nitrogen tests (Briški et al., 1996) which for many years were the only analyses for the 2nd and 3rd top dressing in winter wheat. The finding was as follows:

The analyses of soil Nmin and additional rates of N effects on winter wheat grain yields showed different N target values based on the regression curve for the 1st top dressing for maximum yields and for lower yields. The maximum yields (8.5 t ha^{-1}) were established at 220 to 260 kg ha^{-1} of available N (Nmin plus added nitrogen), but the yields from 6 to 7 t ha^{-1} were achieved at 110 to 140 kg ha^{-1} of available N (R: 0.40). It was concluded that the target value for the 1st top dressing for maximum yields needs to be around 150 kg N ha^{-1}, if there is more available nitrogen in the soil. Nmin rests after harvest on control plots and using 120 kg N ha^{-1} were 121 and 120 kg ha^{-1}, respectively, but only 40 to 48 kg of residual NO_3-N ha^{-1}, which is an appropriate value for water protected areas (Bavec,,1999).

According to the balance of environmental and agronomical impacts, advised recommendation for the 1st top dressing is as follows: N rate kg N ha^{-1} = 110 to 140 kg N ha^{-1} – kg Nmin (from 0 to 0.9 m soil depth).

Investigation of chlorophyll meter values of 13 winter wheat cultivars suggests that readings depend on cultivar (genotype), growth stage and yearly effects. Before 2nd top dressing at stage EC 31/32 the average value varied from 445 to 568, and before 3rd top dressing at stage EC 45/50 it varied from 487 to 580. At stage EC 45/50 the statistically identical values of six cultivars varied from 515 to 553, and the rest of the extreme readings were 487 or 580. At stage EC 31/32 there was no strong correlation ($r=0.134$) between chlorophyll meter readings and grain yield, but there was a stronger correlation between the chlorophyll meter values and grain yield ($r=0.538$, sig. at the 0.01 level) at stage EC 45/50. On the basis of results at stage EC 45/50 only, it is possible to explain the 37% effect of chlorophyll meter readings on grain yield with a quadratic or cubic regression curve (Bavec and Bavec, 2001, online 2006).

The use of chlorophyll meter readings was not in accordance with Wolring (1996) suggestions, because of problematic calibration for varieties and weak correlations of chlorophyll meter readings and grain yield and protein content in the grains. Also based on the project (Bavec, 1999-2001), it was not possible to establish clear guidelines for advisors. However we are collecting more data for establishing more appropriate recommendations for using chlorophyll meter readings in Slovenia.

(iii) If we compare just a few basic reviews from other countries, such as Germany (Olfs et al., 2005), we find that many soil-based methods have been developed to measure the soil mineral N available to plants at a given sampling date. Soil sampling at the start of the growing period and analyzing the amount of NO_3-N (and NH_4-N) is a widespread approach in Europe and the USA. Strategies in the USA for reducing NO_3^- loss through drainage include using nitrate soil tests, improved timing of N application at appropriate rates, and plant monitoring, diversifying crop rotations, using cover crops, reducing tillage, optimizing N application techniques, and using nitrification inhibitors (Dinnes et al., 2002). Due to technical procedures Olfs et al. (2005) suggested that based on data from field calibrations, the soil N pool is filled up with fertilizer N to a recommended amount. Depending on the pre-crop, use of organic manure, or soil characteristics, the recommendation might be modified (±10–50 kg N ha^{-1}). Another set of soil methods has been established to estimate the amount of N that is mineralized from organic soil matter, plant residues, and/or organic manure. Plant-analytical procedures cover the whole range from quantitative laboratory analysis to semi-quantitative "quick" tests carried out in the field. The main idea is that the plant itself is the best indicator for the N supply from any source within the growth period. In-field methods like the nitrate plant sap/petiole test and chlorophyll measurements with hand-held devices or *via* remote sensing are regarded as the most promising, because with these methods an adequate adjustment of the N-fertilizer application strategy within the season is feasible.

Extensive and complex data of plant nitrogen tests in winter wheat in France (Justes et al., 1994), with a combination of soil and plant analyses, were named the JUBIL method (Justes et al., 1994).

As reported by Wood et al. (1992) the field chlorophyll measurements in maize were highly correlated with N tissue concentrations at both growth stages during both years of the study (Wood et al., 1992). But, comparing chlorophyll meter readings and correlations with grain yield in maize showed both hybrid specific (Bavec, 2001) and site-year specific (Ziadi et al., 2008). However, authors doubt that chlorophyll meter indicators are clear indicators for maize N status.

The Use of Application Tools in Slovenia vs. Other Countries

The use of decision support tools by farmers and advisors show discrepancies between basic findings and advice, and the farmer's application. As reported by Olfs et al. (2005) in the huge range of methods proposed so far, the simple mild extraction procedures have gained the most interest, but introduction into practical recommendation schemes has been rather limited. Also the previous question, which supports decision tools for the environmental management of nitrogen (Meynard et al., 2002), is generally unanswered and depends on the development of the country. In theory numerous possibilities exist from direct practical methods (Dinnes et al., 2002) and model based decision support (Meynard et al., 2002). The fact is that non-point loss of N from commercial N fertilizer and decreased demand from the fields to water resources is not caused by any single factor. More likely, the age of crops in crop rotations, with a combination of factors, including tillage, drainage, and a general substitution of purchased N for biological crop selection, organic soil matter levels, hydrology, and temperature (Dinnes et al., 2002) all have significant effects on fertilization efficiency and loss of nitrates into the soil surface. For those in the USA, suggestions were made (Dinnes et al., 2002) to improve monitoring of soil nitrogen for split nitrogen application programs, manage variable rate nitrogen application models and methodologies, develop perennial cash-crop systems, cover crop options and management strategies, and nitrate removal strategies. As a strategy for the USA (Dinnes et al., 2002), in Switzerland and some parts of Germany, permanent control of the water table during the growing season by drainage lines and lysimeters were established more than 25 years ago. Also in Austria, Lithuania, Latvia, Lithuania and Poland, state monitoring of Nmin of some reference locations is performed, which serves as an advisory for nitrogen fertilization for winter cereals in surrounding areas. However, these technologies are usable for flat or slightly sloped fields and are economically limited because of existing field tile drainage lines and organization of the system at the country level. It is not possible to establish this kind of system in less developed countries and regions mainly due to economical reasons. In

Slovenia and most EU eastern countries this kind of system is not established (except very few lysimeters). According to the EU common agriculture policy, good agriculture practice and also nitrogen management supports direct or/and indirect measures depending on each country and usually the EU nitrate directive towards soil sampling to support good agriculture practice in this scope is not clear.

In Slovenia and Slovakia farmers take soil samples by themselves and take them to the lab where lab Nmin analysis is performed. In contrast, some countries like Hungary provide soil N analyses only partially, and Estonia and some Balkan countries do not use any analyses at farm level.

For example, US guidelines encourage use of other nutrient management technologies such as stabilizers, slow release fertilizers, incorporation or injection of organic manures, soil nitrate testing, and other technologies that minimize loss to surface or ground water resources and other supporting practices like: tile line DE nitrification bioreactor, constructed wetland, conservation stream buffer and fall cover cropping system (2013, Code practice for nitrogen fertilization).

If we compare the N approach, information sources and new data in developed countries, like the UK, with its improved farming practices with knowledge transferred to growers (Richards, 2007), with Slovenian activities, then large discrepancies would occur.

In countries like Slovenia, nitrogen supply needs more attention especially if we compare it to the UK proposal for their development, which is the best mirror for further development in less developed systems of knowledge and thinking about practical applications. Again, as we extensively cited, their (Richards, 2007) main tasks are:

- The different methods for quantifying soil nitrogen supply, by estimation, measurement, or both, must be validated and compared. The relative contributions of soil mineral nitrogen, nitrogen mineralized during spring and nitrogen taken up by the crop over the winter, or by maize must be clarified. Guidance is needed for the choice of method for different circumstances taking into account the cost and degree of accuracy to be expected.
- A method is needed for monitoring or modeling seasonal effects on soil nitrogen supply and for providing timely guidance on their impact on fertilizer recommendations.
- The extent to which soil nitrogen is utilized by crops affects nitrogen use efficiency and is a key component of many recommendation

systems. Factors that affect the utilization of soil nitrogen should be identified with a view towards improving nitrogen use efficiency. The assumption that soil mineral nitrogen is recovered by the crop with 100% efficiency (around 60% is the corresponding figure for applied inorganic nitrogen) requires validation for different agronomic conditions. (No experience with the models such as SUNDIAL).

- The need for current protein specifications for bread making wheat (usually minimum 13%) should be reviewed. Developments in varieties, in bread making techniques or in market requirements might allow the use of lower protein grain and smaller nitrogen applications.
- Improving the nitrogen economy.
- The biological basis for grain protein concentration as a retrospective indicator of nitrogen supply must be established. Any other practical indicators must be identified for both cereals and oilseed rape. Once methods are established, guidance for growers in their use is needed.
- Guidance is needed on the relative benefits of standard values and chemical analysis in estimating the nitrogen in livestock manures. Where analysis is preferred, guidance is also needed on sampling methods and on the interpretation of analytical reports. Actions in this area must be coordinated with developments in appropriate software.
- The genetic potential of crops for improved nitrogen uptake and utilization should be explored.
- Growers need greater awareness of the environmental issues associated with nitrates. Clear guidance is needed on identification of high-risk fields and farming practices and on practical mitigation methods.

Similar tasks needs to be discussed and provided elsewhere, where similar protocols were not taken into account.

In the UK more than 200 licensed advisors exist. In other EU countries, especially Eastern European countries and also in Slovenia, there are no special licensed advisers for N management. Most advisers do not read and study N fertilizer requirements.

Environmetal Impacts

As we mentioned, in European countries like Switzerland and some regions of Germany, Austria, Lithuania, Latvia, Lithuania, etc., state monitoring of Nmin as a basis for nitrogen fertilization advising is connected with analyses of the nitrate level in groundwater. Most use of separate, more extensive monitoring, according to the Nitrate directive, which is on an area or regional basis, cover the main cash crops N management only indirectly.

As reported for Slovenian Prekmurje region based on 5,057 mineral nitrogen (Nmin) analyses of field soil from 2006 to 2009, the Nmin content (nitrate and ammonium N) was below 50 kg ha^{-1} N in 53% of the samples. In 16% of samples it was 50-100 kg ha^{-1} N, 17% of samples contained 100-200 kg/ha N, and 14% contained more than the 200 kg ha^{-1} N. In winter the content of Nmin was less than 50 kg Nmin ha^{-1}. In summer months higher values of Nmin were measured in maize (130 kg Nmin ha^{-1}), and lower values in cereals (just 15 kg Nmin ha^{-1}). The main problem was that some farmers did not follow economical and environmental targets; they followed only the rules sufficient for subsidies.

If we compare the data from Nitrate directive reports it is only possible to indirectly conclude the influence of agricultural management on water quality; it is impossible to establish conclusions regarding these two parameters for main cash crops.

In general the EU Commission's report for the period 2004-2007 reveals that 15% of groundwater monitoring stations in the EU-27 found nitrate levels above the limit of 50 mg of nitrates per liter. For Bulgaria, Cyprus, Estonia and Hungary, 91% of monitoring sites found steady or decreasing levels. Austria, Denmark, Finland, Germany, Ireland, Lithuania, Luxembourg, Malta, the Netherlands and Slovenia decided to provide the same level of protection to their whole territory, rather than designate nitrate-vulnerable zones.

Member States had to establish codes of good practice for farmers, to be implemented on a voluntary basis throughout their territory, and develop specific action programs for compulsory implementation by farmers located in nitrate-vulnerable zones (2013 Nitrate directive report). Possibilities, like Soil and Water Assessment Tool (SWAT) for nitrogen (Neitsh et al., 2002; Pohlert et al., 2007) assessment, until now, was not studied in Slovenia in the case of nitrogen (just phosphorus) and it needs a long time for implementation. However, the SWAT model is very promising, because it includes many factors, like algorithms for decomposition, growth of nitrifying bacteria, nitrification, nitrificatory as well as denitrificatory N-emissions, N-uptake by plants and N transport due to water fluxes.

When we follow coordination of the Rural development program 2014 - 2020 in Slovenia, we did not find any positive coordination between agricultural and N environmental measures at the first stage. According to our suggestion Nmin analyses were taken into account as a production measure.

CONCLUSION

Differences and limitations of decision support tools of the nitrogen fertilization system with respect to the implementation of economical and environmental constraints depend upon relationships of findings and knowledge at all levels from scientists to advisers, governmental administration and farmers.

Soil analyses for the amount of NO_3-N (and NH_4-N) at the start of the growing period or at the stage of cereals for N top dressing and the recommended amount of N to target value is a widespread approach in the USA and the EU. Among countries there exists a big variation, and sometimes relationships with water protection is very scarce. The use of nitrate plant sap tests, chlorophyll meter readings or remote sensing is an additional (not main) adjustment of the N fertilizer application. In both soil and plant analyses there exist differences due to soil and plant characteristics, climatic circumstances and potential mineralization, immobilization, N losses, etc.

According to the situation in Slovenia and the known situation in some EU countries, we can conclude that in some countries the environmental priorities of main cash crops (wheat and maize) production management did not follow available findings, and their application at the field level are very scarce. The use of appropriate models is scarce or out of functions, and must be developed and/or adapted. The main problem for using official models is lack of data, such as cost covers.

Because of discrepancies, the minimal N management standards due to production costs and negative environmental impacts must be regulated and supported around the world by the FAO and/or OECD.

According to UK experiences, special tools like SWAT and licenses for advisers for N management must be developed in Slovenia and also in less developed countries.

REFERENCES

Annon. Code practice for nitrogen fertilization. 2013., http://www.acwa-rrws.org/initiatives/cop.html, accessed 20.05.2014.

Annon., 2014. Guidelines for integrated crop management. Ministry of Agriculture and environment, http://www.mko.gov.si/fileadmin/mko.gov.si/pageuploads/podrocja/Kmetijstvo/Integrirana_pridelava/TN_poljscine_2014.pdf, accessed 27.05.2014.

Bavec, F. 1992. Plant population densities, way of sowing and nitrogen as effected to leaf area, elements accumulation and yield of maize hybrids (*Zea mays* L.) FAO groups 100-400. Ph. D. Diss., University of Novi Sad, Faculty of Agriculture (Serbian with English Abstract and Summary), 115 p.

Bavec, F. 1994[a]. Fertilisation of maize with nitrogen on analyses base. Symposium, New challenges in field crop production (Slovenian), Kočevje.

Bavec, F. 1994[b]. Nitrogen (soil N min and fertilizing) effects on maize growth in Slovenian conditions. In: Third congress of the European society for agronomy, Padova University Proceedings. Cedex: European society for agronomy, 1994, str. 446-447.

Bavec, F. 1999-2001. Winter wheat fertilisation with nitrogen. National project (Slovenian), V4- 0308.

Bavec, F., Bavec, M. 2001. Chlorophyll meter readings of winter wheat cultivars in grain yield prediction. Commun. in soil sci. and plant anal. 32: 2709-2719.

Bavec, F., Bavec, M., Fekonja, M. 2013. Organic and mineral nitrogen fertilizers in sweet maize (*Zea mays* L. *saccharata* Sturt.) production under temperate climate. *Žemdirbyste*. 100 (3): 243-250.

Bavec, M. 1999. Yield, yield components and quality of winter wheat grains depending on methods and quantity of nitrogen in Slovenian Podravje. Ph. D. Diss., University of Zagreb, Faculty of Agriculture (Croatian), 209 p.

Briški, L. 1996. Technological and environmental optimization of crops plant nutrition with nitrogen. National project (Slovenian), L4-6554-401-1994/95.

Čabilovski, R. R., Manojlović, M. S., Bogdanović, D., Rodić, V., M. Bavec. 2011. Fertilization economy in organic lettuce production. *Jour. of sust. agric*. 35: 745-756.

Dinnes, D.L., Karlen, D.L., Jaynes, D.B., Kaspar, T.C., Hatfield, J.L., Colvin, T.S., and Cambardella, C.A. 2002. Review and interpretation: Nitrogen management strategies to reduce nitrate leaching in tile-drained Midwestern soils. *Agron, Jour*. 94: 153-173.

European Union the Nitrates directive - Council Directive No. 91/676/EEC.

European Union the Water Framework Directive No. 2000/60/EG, 2000.

Justes. E., Meynard, J-M., Mary, B., Laurent, F. 1994. JUBIL: A new method of conducting the nitrogen fertilization in winter wheat crops. Proc. 3rd ESA Congr, Padova, pp 490-491.

Latković, D., Marinković, B., Crnobarac, J., Bogdanović, D., Bavec, F., Jocković, Đ. 2010. Distribution of NO_3-N within the soil profile in enhancing corn yield. *Savremena poljoprivreda*. 2010, 59: 355-362.

Meynard, J-M., Cerf, M., Guichard, L., Jeuffroy, M-H., Makowski, D. 2002. Which decision support tools for the environmental management of nitrogen? *Agronomie* 22: 817–829.

Neitsch, S.L., Arnold, J.G., Kinry, J.R., Williams, J.R., 2002. Soil and Water Assessment Tool. Version 2000. Theoretical Documentation. *USDA-ARS*, Temple, TX.

Olfs, H.-W., Blankenau, K., Brentrup, F., Jasper, J., Link, A., Lammel, J. 2005. Soil- and plant-based nitrogen-fertilizer recommendations in arable farming. *J. Plant Nutr. Soil Sci*. 168: 414–431.

Regulation (EC) No. 1137/2008 of the European Parliament and of the Council of 22 October 2008.

Regulation (EC) No. 1882/2003 of the European Parliament and of the Council of 29 September 2003.

T. Pohlert,T., Huisman, J.A., Breuer, L., Frede, H.-G. 2007. Integration of a detailed biogeochemical model into SWAT for improved nitrogen predictions—Model development, sensitivity, and GLUE analysis. *Ecological modelling*, 203: 215–228.

Wollring, J. 1996. Gezielte Schosser- und Aehrengabe mit dem Hydro N-tester: N Bedarfsermittlung per fotometrischer Pflanzenanalyze. Jour., 18-20.

Wood, C. W., Reeves, D. W., R.R. Duffield, R. R., and Edmisten, K.L. 1992. Field chlorophyll measurements for evaluation of corn nitrogen status. *Jour. Plant. Nutr*. 15: 487-500.

Ziadi, N., Brassard, M., Belanger, G., Claessens, A., Tremblay, N., Camborius, A.N., Nolin, M.C., Parent, L-E. 2008. Chlorophyll Measurements and Nitrogen Nutrition Index for the Evaluation of Corn Nitrogen Status. Agron. Jour., 100.

INDEX

#

A

B

C

F

G

H

Q

R

U

V

W

Y